GENETICS - RESEARCH AND ISSUES

MATHEMATICAL GENETICS

GENETICS - RESEARCH AND ISSUES

Additional books in this series can be found on Nova's website under the Series tab.

Additional e-books in this series can be found on Nova's website under the e-book tab.

GENETICS - RESEARCH AND ISSUES

MATHEMATICAL GENETICS

ANDREY NIKOLAEVICH VOLOBUEV

NOVA BIOMEDICAL

New York

Library of Congress Cataloging-in-Publication Data

ISBN: 978-1-63463-254-6
Library of Congress Control Number: 2014954117

Published by Nova Science Publishers, Inc. † New York

CONTENTS

PREFACE

In this book mathematical aspects of population genetics are considered.

On the basis of the Hardy-Weinberg law the standard approach to the research of population genetics problems is stated. Along with the standard approach the necessity for separate research of family tree genetics and population genetics which represent a set of the family trees is shown.

Family trees are investigated by methods of discrete mathematics in a discrete time scale which is defined by the alternation of generations. In the research of populations it is necessary to transit to a continuous time scale, continuous functions; therefore, the Hardy-Weinberg law is written down in the form of the differential equation of the second order. Transition to continuous functions has allowed us to receive new and significant results in population genetics. In particular, new results affect the approach to problems of a mutation's occurrence under action of radiation, of a newgrowth's occurrence, migrations of populations under various conditions, to reveal nonlinear character of inbreeding and natural selection, etc.

The book may be useful to geneticists, student-biologists, post-graduate students and anyone who is interested in problems of population genetics.

Chapter 1

1. INTRODUCTION

Mathematical genetics is a part of genetics which aims at the description of a set of live reproductive individuals based on methods of mathematical analysis.

In mathematical genetics, as well as in any mathematical areas of a science, objects of research have an idealized character. At the analysis alongside key parameters of object, only such properties and characteristics of the object which will be interesting to the purposes of research are allocated.

The set of live reproductive individuals can be divided into two types for which mathematical analysis is carried out differently.

The first type of research object is *a family tree*.

It is supposed that during any conditional moment of time the man and the woman - founders of the family tree - give posterity. This posterity is referred to as the first generation. Further descendants in turn give posterity, which is referred to as the second generation, etc. Usually, individuals of the given generation choose a partner for reproduction from other family trees. It is connected to experimental supervision which shows that a similar choice of the partner gives better posterity. However, the choice of a closely related partner and from the family tree - so-called *inbreeding* - is possible.

For the analysis of a family tree the methods of discrete mathematics and probability theory of random events, etc. are used.

The second type of object is *a population*.

The population will usually consist of plenty of family trees.

Consider *panmictic* (from the Greek *pan* - general and *mixis* - mixing) populations and *inbred* populations.

The panmictic population first of all, is characterized by the absence of consanguineous marriages - inbreeding. In reality, a population is believed panmictic if during the analysis it is possible to neglect the inbreeding. In a family tree a marriage is not considered as consanguineous if in the last three generations of the marrying individuals there are none who are closely related. Panmixia of populations are reduced also with selection which at reproduction prefers individuals with certain characteristics.

For inbred populations the inbreeding plays an essential role and it needs to be taken into account in the analysis. The inbred populations usually represent a closed set of individuals owing to geographical, ethnic, class, religious or isolation issues, etc.

For the analysis of populations, methods of continuous functions are used, differential equations, etc.

Transition to continuous functions has allowed us to receive new and significant results in population genetics. In particular, we are able to consider the problem of a migrating population under various conditions, to reveal the nonlinear character of inbreeding and natural selection, and take a new approach to the phenomenon of crossing-over, etc.

This book may be useful to geneticists, students-biologists, post-graduate students and anyone who is interested in problems of population genetics.

Chapter 2

2. Mathematical Genetics of a Family Tree

A family tree is essentially a discrete object of research. The family tree represents consistently alternating generations. Simultaneously, there can be several generations; 3 - 4 generations is usual. The average time of a human individual lifespan is frequently rather long: 70 - 90 years. But the period of alternating generations, i.e., the occurrence of a new generation in the presence of the previous generation, is much less and is $T = 25 - 30$ years.

Let's note that the changes which take place with an individual during his life, i.e., ontogenesis after the end of the reproductive period, are not of interest to us. All genetic attributes which can be transferred to posterity should be present in an individual up to reproduction, i.e., until the time that the generation alternates.

If any attribute of an individual is not fixed in its genome, this attribute is not examined at the analysis either; for example, an infectious disease, a trauma, etc.

2.1. The Hardy: Weinberg Law for a Family Tree

In the basis of mathematical genetics one law was discovered in 1908 simultaneously by an English mathematician, Hardy, and a German physician, Weinberg. These scientists based their research on earlier experiments which were carried out by Gregor Mendel.

2.1.1. Mendel's Experiments

The first indications that mathematical methods can play a big role in biology came from the Austrian (Czech) scientist G. Mendel. He published his conclusions in 1865 in "Verhandlungen des Naturforschenden Vereins" (Brunn).

Mendel carried out experiments on various sorts of pea plants (Pisum sativum) which differed in a line of attributes: color of seeds - yellow and green, shape of seeds - round and angular.

He found out that if one was to cross peas to yellow and green seeds in the first generation, the received plants would have only yellow seeds.

In the second generation with parental yellow seeds there would be a splitting color attributes of seeds. Plants have both yellow and green seeds. Thus, their ratio equals 3:1 for the benefit of yellow seeds.

Complicating the experiments, Mendel used peas with yellow and round seeds and with green and angular seeds. Thus, in the first generation there were only yellow round seeds. In the second generation it appeared that there was a splitting of attributes on color and on shape independently from each other. Color splitting occurred in the ratio of 3:1 for the benefit of yellow seeds, and shape in the ratio of 3:1 for the benefit of round seeds. Hence, both splittings of attributes have the same ratio. Splitting of both kinds of attributes occurs independently in the ratio $(3:1)(3:1)=9:3:3:1$ - yellow round seeds, yellow angular, green round and green angular seeds.

Let us note that despite the use of the mathematical statistics methods in the analysis of results of the experiments, the received laws do not carry statistical character, but a determined character, i.e., they are related to base laws of biology.

Mendel's experiments have allowed us to begin to create a terminology of genetics.

Attributes which prevail after mating began to be called *dominant*; for example, the yellow color of pea seeds, or the roundness of their surface.

Attributes which are suppressed after mating are referred to as *recessive*; for example, the green color of pea seeds, or the angularity of their surface.

The set of all attributes of the individual which are found by visual supervision or in laboratory researches is called *the phenotype* of an individual. For example, the color of the pea seeds and the condition of their surface is concerned with phenotypic attributes.

Mendel offered a model of discrete *genes* which control the given attributes. He also assumed that one gene controls one attribute.

As attributes are dominant and recessive, it is logical to assume that each gene has at least two forms.

Allele (from the Greek *another*) is one of two or more alternative forms of one gene.

Alleles which define the dominant attributes of an individual are referred to as *dominant alleles*. Dominant alleles are frequently designated by the letter A.

Alleles which define the recessive attributes of an individual are referred to as *recessive alleles*. Recessive alleles are frequently designated by the letter a.

Furthermore, the so-called sexual reproduction of organisms will be of interest to us. The sexual (reproductive) cell of an organism (animals or plants) is referred to as *gamete*. For the human: in women it is egg or ovum; in men it is spermatozoon.

Gametes in the nucleus have chromosomes which contain various genes. *A locus* is a site of the given gene on a genetic map of the chromosome. Actually, genes are sites of chromosomes. Chromosomes in gametes usually have only one kind of allele from the given pair: dominant or recessive.

Each kind of plant or animal, including the human, has a certain number of chromosomes in the gametes. In the human there are 23 chromosomes. Of these there are 22 so-called *autosomes* and 1 *sex chromosome*. In the structure of the autosomes between men and women there is no basic distinction. Sex chromosomes can be one of two kinds: *X-chromosomes* or *Y-chromosomes*. Thus, the gametes of women (ovum) have only the *X*-chromosome. Gametes of men (spermatozoon) can have either the *X*-chromosome, or the *Y*-chromosome.

Alleles taking place in sex chromosomes are referred to as being *linked with a sex* (sex-linked) or particularly *linked with the X-chromosome* (X-linked) or with the *Y*-chromosome (Y-linked).

In the sexual act there is probably an association of gametes (ovum and spermatozoon). There is a new cell which is referred to as *a zygote* – fertilized ovum. In the zygote there is a double (diploid) set of chromosomes. If in the zygote two *X*-chromosomes were united, from such a zygote, a female individual will develop. If one *X*-chromosome and one *Y*-chromosome were united, from such a zygote a male will develop. Both processes are practically equiprobable. The given conclusion can be proven following Table 1 of individual different sexes mating.

When mating occurs between the man m, whose gametes contain either the *X*-chromosome or the *Y*-chromosome, and the woman f, whose gametes contain only the X-chromosome, the male and female zygotes are formed in the ratio 1:1. Factors before a designation of sex chromosomes reflect fractions of these chromosomes in gametes and zygotes of a population.

Table 1.

$f \backslash m$	$\frac{1}{2}X$	$\frac{1}{2}Y$
$\frac{1}{2}X$	$\frac{1}{4}XX$	$\frac{1}{4}XY$
$\frac{1}{2}X$	$\frac{1}{4}XX$	$\frac{1}{4}XY$
Σ	$\frac{1}{2}XX$	$\frac{1}{2}XY$

The zygote always has the certain *genotype* to the given (examined) attribute (property). At the association of gametes a zygote occurrence of two dominant alleles is possible. In this case there is genotype AA or a cell - *dominant homozygote* to a considered attribute. At the association of two recessive alleles there is a genotype aa or *recessive homozygote*. If there are associated dominant and recessive alleles, there is genotype Aa or *heterozygote*.

The possible combinations of alleles at the association of gametes in zygotes and genotypes are shown in Table 2.

Table 2.

$f \backslash m$	A	a
A	AA	Aa
a	Aa	aa

Thus, at mating the man *m*, whose gametes contain either allele *A* or allele *a*, and the woman *f*, with the same possible set of alleles to the considered attribute occurrence of posterity of 3 various genotypes is possible: *AA*, *Aa*, *aa*. Taking into account that allele *A* is dominant and the allele *a* is recessive, there are 2 phenotypes. *The dominant phenotype* is formed if there are genotypes *AA* and *Aa*. *The recessive phenotype aa* is formed at the crossing of the recessive gametes. From Table 2 it can be seen that at 3 opportunities of the dominant phenotype occurrence there is 1 opportunity of the recessive phenotype occurrence. In this the essence of Mendel's law of the attributes splitting 3:1 will consist.

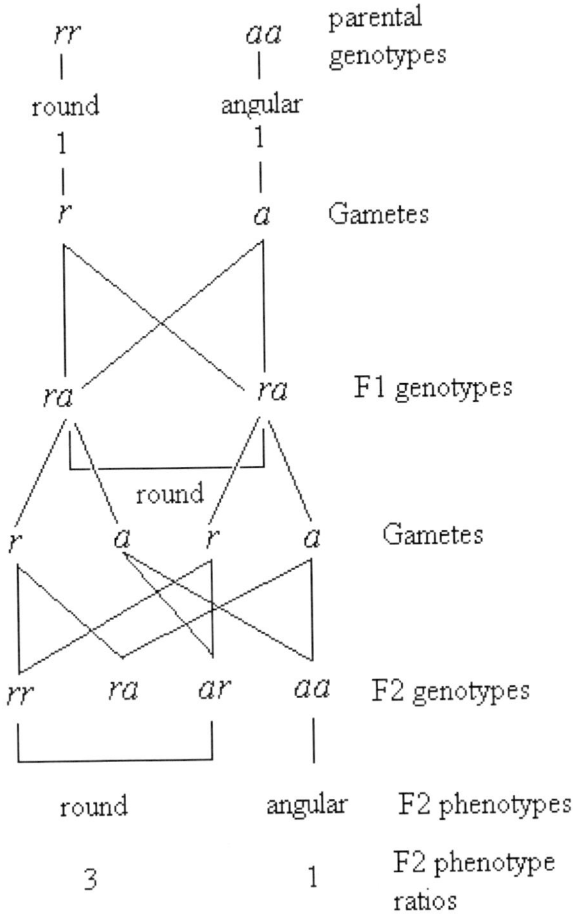

Figure 2.1.

In Figure 2.1 the diagram which illustrates Mendel's experiments on crossing peas with round and angular seeds in the ratio 1:1 [23] is shown.

In Figure 2.1 the numbers of pea generations there are designated F1 and F2. Parental genotypes *rr* – dominant homozygotes and *aa* – recessive homozygotes form gametes with dominant alleles *r* and recessive alleles *a*. These gametes being crossed give first generation F1 with round (genotype *ra* - heterozygota) seeds. Genotypes of first generation F1 form in turn gametes with alleles *r* and *a*. These gametes being crossed give second generation F2 with round (genotype *rr* – dominant homozygote, *ra*, *ar* - heterozygotes) and angular (genotype *aa* – recessive homozygote) seeds in the ratio 3:1.

2.1.2. The Balance of Hardy-Weinberg

Let us consider the essence of the Hardy-Weinberg law.

We assume that in a family tree at the examined individuals in the autosomal genome in gametes the frequency of the dominant alleles A equals p, and the frequency of the recessive alleles a equals q. Thus, the sum of frequencies is equal to unit $(p+q=1)$.

The frequency, for example, of autosomal dominant allele A equals $p=\frac{m}{n}$, where n is the general number of family tree individuals of the given generation, m – is the number of individuals who have dominant allele A. As other individuals have recessive allele a the sum of frequencies dominant and recessive alleles is equal to unit. At examination of the autosomal genome the sexual distinction of gametes does not play a role. Both allele A and allele a can have male and female individuals.

The autosomal genome defines in particular the size of the organs. For example, there is a feature of an organism called the brachydactylia (in Greek, *short finger*), in which the second bone in the finger is shorter than normal. In such people not only the phalanxes of the fingers and toes are short but also it may happen that the number of phalanxes is reduced. The given feature is inherited on autosomal-dominant type [3]. The reproductive function in people with brachydactylia is not broken.

Let us write down the ratio of alleles frequencies as:

$$(A)\,p:(a)\,q. \tag{2.1}$$

At mating of the individuals with different sexes in zygotes there is a doubling of autosomes. In the following generation for autosomal genome (for genes not taking place in sex chromosomes) the frequency balance or the allele ratio – the so-called Hardy-Weinberg balance - is established. This balance by Hardy-Weinberg law is defined.

According to the Hardy-Weinberg law, the genotypes *AA*, *Aa* and *aa* in zygotes have the following frequencies:

$$(AA)\,p^2 : (Aa)\,2pq : (aa)\,q^2, \qquad (2.2)$$

The frequency ratio (2.2) of genotypes in individual zygotes can be received using the expansion of a square binomial:

$$(p+q)^2 = p^2 + 2pq + q^2 = 1. \qquad (2.3)$$

The essence of the ratio (2.2) will be that frequency p^2 of genotype *AA* arising in the fertilized zygote is equal to the product of probability *p* of receiving the dominant allele *A* from the mother and the probability *p* of receiving the dominant allele *A* from the father, i.e., $p \cdot p = p^2$. The same is true for a genotype *aa* which has probability of occurrence in a zygote $q \cdot q = q^2$. For genotype *Aa* the situation is different. It can arise in a zygote if the gamete of the mother has the allele *A* with the probability *p* and the gamete of the father has the allele *a* with probability *q*. Or another variant is possible: the gamete of the mother has the allele *a* with probability *q* and the gamete of the father has the allele *A* with probability *p*. Therefore, the general probability of occurrence in the zygote of genotype *Aa* equals $p \cdot q + q \cdot p = 2pq$.

As was already specified earlier, the dominant alleles define a phenotype of an individual i.e., including in the laboratory attributes; for example, the color of eyes, hair, growth, group of blood, etc. In the ratio (2.2) three genotypes are present but only two phenotypes; the first phenotype is defined by a combination of genotypes $AA + Aa$, and the second phenotype by one genotype *aa*.

The ratios (2.1) and (2.2) remain constant in a family tree from generation to generation only in the case of an ideal population (where the number of individuals is very great, panmixia exists, and there is no selection, mutations, or migrations of individuals, etc.) [1].

The major property of the Hardy-Weinberg balance is that it will be *neutral.*

For autosomal inheritance of allele frequencies in gametes it is obvious enough. Using distribution of genotypes (2.2), it is possible to receive, for example, the frequency of the recessive alleles a in the following $(n+1)$ generation. For this purpose it is necessary to do summation for the previous generation n for the half of heterozygotes frequency Aa and the homozygotes frequency aa:

$$q_{n+1} = \frac{1}{2} 2 p_n q_n + q_n^2 = q_n (p_n + q_n) = q_n . \tag{2.4}$$

In the following generation the same frequency of allele a as in the previous explanation is received.

The rule for finding the allele frequency in the following generation is one we shall use on a regular basis; therefore, we shall analyze it in more detail.

If there are gametes for men and women with the allele ratio (2.1) at their crossing, there are zygotes of the following generation with ratio of genotypes (2.2). This ratio of genotypes is shown in Table 3, similar to Table 2, where instead of alleles and genotypes their frequencies are shown.

Table 3.

f \ m	p	q
p	p^2	pq
q	pq	q^2

From these zygotes individuals of the following generation who have gametes of this generation develop. For finding the allele frequencies in these gametes it is necessary to sum the frequencies of genotypes either on lines or on columns with the account $p+q=1$. In it the essence of the formula (2.4) will consist.

Disturbance of the ratio (2.2) due to any external influences - mutagenesis, migrations, etc. - are resulted only in change of the genes frequency ratio (2.1) in the following generation and to restoration of the ratio (2.2).

Influences on a population can be connected to migrations, occurrence of subpopulations, an electromagnetic and radiating irradiation, etc. For a human population an essential influence can also be wars, negative public phenomena connected with the destruction of classes and public groups of the population.

The mechanical realization of three possible types of balance (stable - 1, unstable - 2, indifferent – 3) is shown in Figure 2.2.

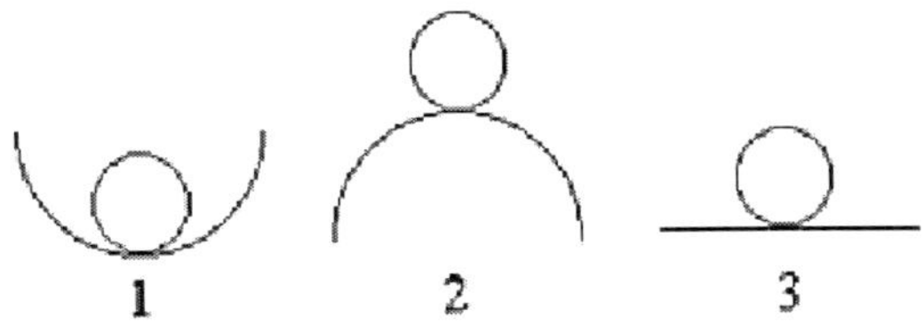

Figure 2.2.

At influence on a population - for example, mutagen factors of the various kinds - the Hardy-Weinberg balance is restored; however, the population does not compensate for the influence. The allele frequencies change.

The indifferent character of the Hardy-Weinberg balance results in the occurrence of the external influence leading to the deterioration of a population quality which cannot be compensated by the population even if this influence has stopped. Restoration of the alleles initial ratio is possible only due to their receipt from the outside i.e., immigrations from other populations.

Let us note that Russian tsars Peter I and Ekaterina II, at the constant invitation of various experts from residences in Western Europe, were actually engaged in the improvement of a genofund of the Russian population.

Let us note also that the Hardy-Weinberg law plays the same basic role in biology as the second Newton's law in mechanics. However, in the attitude of the world around these two laws of genetics and mechanics there is an important distinction.

The second law of Newton has the same form in the Universe.

There is another situation for the Hardy-Weinberg law. The exponent 2 in the binomial (2.3) specifies that two different sexes of the individuals are necessary for reception of posterity. In people there are males and females. This fact reflects the concrete conditions of a life occurrence on the Earth. It is possible to assume that on other planets this exponent will have another value.

For example, for reception of posterity in an aggressive enough environment it might be necessary to have 3 individuals of a various sex. In particular 2 different sexes of the individuals at mating should be located inside of the third individual; genetically contacting with it anyway. In this case the exponent in the formula (2.3) should equal 3.

2.1.3. Genetic Base of ABO Blood Groups System

Let us consider the inheritance of blood groups in humans. In spite of the fact that the gene of blood is linked to the *X*-chromosome, the frequencies of genic alleles in men and women are identical. Therefore, linking with a sex chromosome in the absence of the mutagen factor working on blood can not be taken into account to formally analyze the inheritance of the blood genotypes as autosomal. Crossing-over apparently also does not affect the blood alleles, see paragraph 2.1.5.

The problem of blood classification has arisen as a result of the necessity for blood transfusion after significant blood loss; for example, in the case of a wound.

First, a blood transfusion from a male to a male was carried out in England in 1819 by the professor of obstetrics and gynecology J. Blundell. But it was only in the beginning of the twentieth century that the Austrian pathologist K. Landsteiner established that the blood of people can be divided into 4 groups on biological properties - 4 phenotypes; for this he was awarded the Nobel Prize.

In the further analysis we shall follow the German geneticist Felix Bernstein who achieved it in 1925 [2, 3].

If the blood of any group combined with the blood of the group incompatible with it there is a pasting (agglutination) of erythrocytes, their destruction.

Blood which can be transfused to all people is the group I (O).

Blood which is compatible with the blood of group IV is group II (A).

Blood which is also compatible with the blood of group IV is group III (B).

However, the blood groups II and III are incompatible with each other.

Blood which can be transfused only to people with the same group of blood is the group IV (AB).

Groups of blood are inherited and do not vary during a lifespan.

In Table 4 the distribution of relative frequencies of blood groups in Berlin in 1925 is shown.

Table 4.

Blood groups	O	A	B	AB
Frequencies	0.366	0.4323	0.1415	0.0601

Bernstein in his analysis used the three-allele model of a blood gene.

There are equilibrium frequencies of genotypes p, q, and r in a locus with three alleles A, B and O so $p+q+r=1$.

According to the Hardy-Weinberg law the distribution of the genotypes frequencies looks similar to decomposition $(p+q+r)^2$:

$$p^2(\mathrm{AA})+q^2(\mathrm{BB})+r^2(\mathrm{OO})+2pq(\mathrm{AB})+2pr(\mathrm{AO})+2qr(\mathrm{BO}) \tag{2.5}$$

Hence, there are all 6 genotypes of blood.

Four phenotypes of the blood result from alleles A and B domination above allele O.

In Table 5 according to distribution (2.5) four classes of phenotypes with their frequencies following from the Hardy-Weinberg law are shown.

Designating frequencies of the blood phenotypes hyphens on top, we have:

$$\overline{\mathrm{O}}+\overline{\mathrm{A}}=(r+p)^2=(1-q)^2 \text{ and } \overline{\mathrm{O}}+\overline{\mathrm{B}}=(r+q)^2=(1-p)^2. \tag{2.6}$$

Hence, interrelation of the allele frequencies and the blood phenotypes frequencies looks like:

$$q=1-\sqrt{\overline{\mathrm{O}}+\overline{\mathrm{A}}},\ p=1-\sqrt{\overline{\mathrm{O}}+\overline{\mathrm{B}}} \text{ and } r=\sqrt{\overline{\mathrm{O}}}. \tag{2.7}$$

Substituting in formulas (2.7) the frequencies of phenotypes from Table 4, we find the frequencies of the blood gene alleles:

$$p=0.2876;\ q=0.1065;\ r=0.6050. \tag{2.8}$$

Table 5.

Blood groups	I(O)	II(A)	III(B)	IV(AB)
Blood phenotypes	OO	AA+AO	BB+BO	AB
Phenotypes frequencies	$\overline{\text{O}} = r^2$	$\overline{\text{A}} = p^2 + 2pr$	$\overline{\text{B}} = q^2 + 2qr$	$\overline{\text{AB}} = 2pq$

For control we summarize the allele frequencies and receive $p + q + r = 0.9991$ which will be very well coordinated to the expected size 1.

2.1.4. Hardy-Weinberg Law Linked to *X*-Chromosome Inheritance

For examination of the Hardy-Weinberg law for inheritance linked to the *X*-chromosome it is necessary to take into account that as opposed to autosomes in men and women, in the *X*-chromosomes for the account of the crossing-over constant distinction of the allele frequencies is supported.

Let us consider inheritance of a two-allele gene linked with the *X*-chromosome. The frequency of a dominant allele *A* we shall designate in the men p_m and in the women p_f. For recessive allele *a* it is accordingly q_m and q_f.

At mating there is a ratio of genotypes in the women according to product $(p_f + q_f)(p_m + q_m)$. Thus:

$$(AA)\, p_f p_m : (Aa)(p_m q_f + p_f q_m) : (aa)\, q_m q_f \, . \qquad (2.9)$$

The men have *hemizygous* on genes in the *X*-chromosome frequency ratio. Hemizygous refers to the frequency allele ratio in a zygote caused by the absence of the *homologous* (from the Greek *corresponding*) second *X*-chromosome. Instead there is a *Y*-chromosome.

In connection with that the *X*-chromosome of the woman at mating passes to the male offspring; in the zygote of men arises the hemizygous allele ratio:

$$(A)\,p_f : (a)\,q_f \,. \tag{2.10}$$

Let us find the frequency of the recessive allele *a* in the woman in the following generation. For this purpose the rule applied earlier for the autosomal genome is used.

For finding the frequency of recessive alleles *a* in women in a new generation it is necessary to add the frequency of homozygotes *aa* and half of the frequency of heterozygotes *Aa* from a ratio (2.9). It is similarly necessary to do this for alleles *A*. Using this form of calculation of allele frequencies we find a change of alleles *a* frequency in generations of women equally:

$$\begin{aligned} q_{f(n+1)} &= \frac{1}{2}\left(p_{mn}q_{fn} + p_{fn}q_{mn}\right) + q_{mn}q_{fn} = \\ &= \frac{1}{2}\left(p_{mn} + q_{mn}\right)q_{fn} + \frac{1}{2}\left(p_{fn} + q_{fn}\right)q_{mn} = \frac{1}{2}\left(q_{fn} + q_{mn}\right) \end{aligned} . \tag{2.11}$$

At the deduction (2.11) the conditions of the gene two-alleles for men $p_{mn} + q_{mn} = 1$ and for women $p_{fn} + q_{fn} = 1$ are used.

The frequency of the dominant allele *A* in generations of women is calculated under the similar formula:

$$\begin{aligned} p_{f(n+1)} &= 1 - q_{f(n+1)} = 1 - \frac{1}{2}\left(q_{fn} + q_{mn}\right) = \\ &= \frac{1}{2}\left(1 - q_{fn} + 1 - q_{mn}\right) = \frac{1}{2}\left(p_{fn} + p_{mn}\right) \end{aligned} . \tag{2.11a}$$

Hence, for the genes linked to the *X*-chromosome the Hardy-Weinberg balance in the women is established in the following generation as at the autosomal genes.

The frequency of the allele linked with the *X*-chromosome in the following generation is equal to a half-sum of the frequencies of male and female alleles in the previous generation.

It is also necessary to take into account the major condition that allele frequency in the X-chromosome of the women is equal to the frequency of the same alleles in men in the following generation. For recessive allele a this condition looks like:

$$q_{mn} = q_{f(n-1)}. \qquad (2.12)$$

The condition (2.12) is caused by the X-chromosome of the woman at inheritance (mating) passing to the male offspring.

Substituting (2.12) in (2.11), we find:

$$q_{f(n+1)} = \frac{1}{2}\left(q_{fn} + q_{mn}\right) = \frac{1}{2}\left(q_{fn} + q_{f(n-1)}\right). \qquad (2.13)$$

The received recurrent formula (2.13) allows us to understand how there is an approach to the Hardy-Weinberg balance for the genes linked to the X-chromosome. For the solution of the recurrent ratio (2.13) we shall write it down deviated on one generation back:

$$q_{f(n-1)} - 2q_{fn} + q_{f(n-2)} = 0. \qquad (2.14)$$

The solution of the finite-differential equation (2.14) we find as $q_{fn} = a^n$ where in this case a is constant. Substituting this solution in the formula (2.14), we have:

$$a^{n-1} - 2a^n + a^{n-2} = 0. \qquad (2.15)$$

Let us divide the equation (2.15) by a^{n-2}:

$$a - 2a^2 + 1 = 0. \qquad (2.16)$$

We find two roots of the characteristic quadratic (2.16):

$$a_1 = 1 \text{ and } a_2 = -\frac{1}{2}. \tag{2.17}$$

Hence, the general solution of the finite-differential equation (2.14) looks like:

$$q_{fn} = C_1 + C_2\left(-\frac{1}{2}\right)^n. \tag{2.18}$$

We shall find constants of integration C_1 and C_2 on the basis of the initial conditions: at $n = 0$, $q_{fn} = q_{f0}$ and at $n = 1$ according to (2.11) $q_{fn} = q_{f1} = \frac{q_{m0} + q_{f0}}{2}$. Thus:

$$C_1 = \frac{2q_{f0} + q_{m0}}{3} \text{ and } C_2 = \frac{q_{f0} - q_{m0}}{3}. \tag{2.19}$$

Therefore, the solution (2.18) finally looks like:

$$q_{fn} = \frac{2q_{f0} + q_{m0}}{3} + \left(\frac{q_{f0} - q_{m0}}{3}\right)\left(-\frac{1}{2}\right)^n. \tag{2.20}$$

On Figure 2.3 the graph of the frequency q_{fn} of allele *a* linked with the *X*-chromosome in generations of the women is shown. As allele frequencies in women and men initially can be unequal, the initial conditions for calculation are accepted as $q_{f0} = 0.6$ and $q_{m0} = 0.2$.

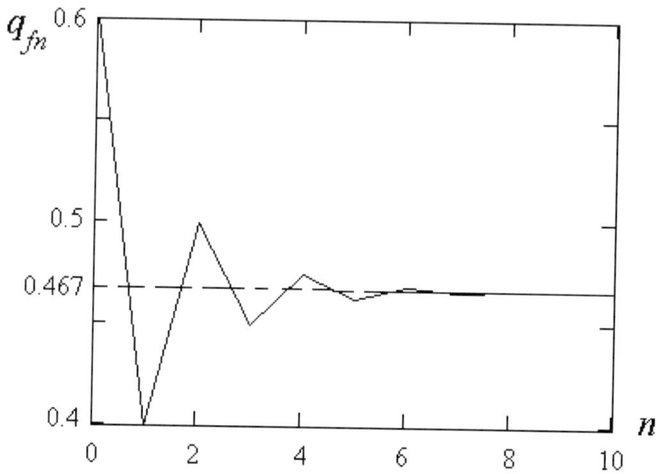

Figure 2.3.

It can be seen from the graph that the frequency of allele a in the X-chromosome of the woman gradually comes nearer to the equilibrium frequency equal according to (2.20) $q_{feq} = \dfrac{2q_{f0} + q_{m0}}{3} \approx 0.467$. This balance is established approximately for n = 8 generations.

It is possible to estimate the velocity which Hardy-Weinberg balance establishes in a family tree for the genes linked to sex chromosomes using:

$$q_{f(n+1)} - q_{m(n+1)} = q_{f(n+1)} - q_{fn} = \frac{1}{2}\left(q_{fn} + q_{mn}\right) - q_{fn} = \frac{1}{2}\left(q_{mn} - q_{fn}\right) \tag{2.20a}$$

Hence, in each generation of a family tree the difference between allele frequencies in the sex chromosomes of men and women decreases twice and finally disappears. In particular, therefore, there is no distinction in allele frequencies which determine the blood genotypes, see paragraph 2.1.3.

2.1.5. Crossing-Over in a Family Tree and Evolution

When examining the Hardy-Weinberg law there are questions. Why are the autosomal genes inherited under the law (2.2) and the sex-linked genes under law (2.9)? What is the biological expediency of such distinction?

Before answering these questions we shall consider why there is a biological variety in a human population. Here it is necessary to pay attention to three positions:

a) All people have more identical attributes than differences. All people have one head, two hands, two legs, five fingers on each hand, five toes on each foot, two eyes, two ears, one heart which is on the left of the thorax, two kidneys, etc. The information on all it should be in autosomes and should be cloned from generation to generation under the law (2.4). The given community of individuals creates population monotony, i.e., actually separates one population of the living world from another; for example, the human from peas.
b) All people differ regarding sex. It is known that the information on sexual attributes is contained in the X and Y chromosomes. It is obvious the purpose of the given distinction will consist in the necessity of the population to achieve reproductive constancy or increase in the number of a population.
c) Individuals of one sex differ from each other a little.

This distinction between different populations may be relatively constant and predicted; for example, the distinction in color seeds of the pea population. The given distinction of attributes connected to the distinction of the dominant and recessive alleles of genes in gametes is transferred from generation to generation in the ratio 3:1 for the benefit of yellow seeds as opposed to green; or round seeds as opposed to angular seeds, as shown in Figure 2.1. But as a whole, despite such distinction, there is simply a cloning of an attribute from generation to generation. It is possible the given ratio of colors of pea seeds or, for example, the flower colors of any plants improve the ability of these plants to reproduce; in particular, the pollination of flowers by insects preferring this or that color. G. Mendel broke the ratio 3:1 crossing of pea seeds in the ratio 1:1, Figure 2.1, which allowed him to find the laws of attribute inheritance.

People also are born with: different race, different color of hair, eyes, different blood groups; they may be taller, shorter, thinner, fatter, more or less intelligent, sillier, stronger, weaker, with a long or short life span, etc. For example, the race, frequency distinction of blood groups and other attributes are transferred from generation to generation.

The allele's kind of attribute distinction is directed toward the optimization and simplification of the individual marriage contacts and also on increase in the competitiveness of individuals of a population in an area of its dwelling.

If we exclude sexual distinction and constant cloned allele's distinction there is still a question regarding the distinction of people.

Obviously, in the genome there must be a way for the constant information to be stored and a way or ways of probabilistic, stochastic change (but in rather small volume) of this information. Such a change of the information is necessary for the evolution of a population and reception of the new useful attributes in individuals.

The ways of information storing of the various kinds (constant and changeable) should be reflected necessarily in the Hardy-Weinberg law.

One way of storage and transfer of the changeable information has already been discussed earlier. Distinction is possible in the alleles of one gene. However, the given distinction is conventional enough. In a family tree the frequency set of genotypes, first of all in autosomes, will be kept from generation to generation, see (2.4). In men and women a set of autosomal alleles is the same. Frequencies of these alleles in men and women are also the same: dominant with frequency p and recessive with frequency q. If frequencies of the men and women autosomal alleles were once various for many generations, the allele frequencies in men and women were equal and invariable repeat what the formula (2.4) testifies.

Another situation regards the X-chromosome. In it the distinction of the dominant and recessive allele frequencies in men and women continuously from generation to generation is observed. Therefore, it is necessary to make a calculation of genotypes inheritance under the formula (2.11). The frequency of dominant and recessive alleles in men and women varies from generation to generation, as shown in Figure 2.3.

However, at first sight there is no distinction between autosomes and X-chromosomes. The frequencies of alleles increase in the number of generation that aspire to the equilibrium value (see Figure 2.3). But in reality, it is not so. There is a mechanism constantly supporting the distinction of the dominant and recessive allele frequencies in men and in women in X-chromosomes. This mechanism is referred to as *crossing-over*. Actually, crossing-over is similar to the continuous disturbance of the allele frequencies in generations shown in G. Mendel's experiments.

At once we shall notice that in the autosomes there is apparently no crossing-over. This specifies the fact established by T. Morgan that in the male of a drosophila (as opposed to the female) the crossing-over is absent [13]. And as in male there are all sets of autosomes, the crossing-over in autosomes is absent. Otherwise the Hardy-Weinberg law would have the same kind in X-chromosomes and autosomes because for many generations the difference of the allele frequencies in men and women should be equal (see the formula (2.20a)).

The *crossing-over* is the exchange of sites, and hence the genes between homologous chromosomes. This exchange usually occurs at meiosis, and is more concrete at *conjugation* when the homologous male and female X-chromosomes closely approach each other (Figure 2.4a) and are interlaced. Thus, a break of the chromosomes threads – *chromatids*, Figure 2.4b – and repeated connection of these threads (but already for two different chromatids, Figure 2.4c) is possible. It results in the transition of the part alleles from a male in female X-chromosome and *vice versa*.

The formula (2.11) specifies that without crossing-over, but in the presence of distinction in allele frequencies in men and women in X-chromosomes, the frequency of the recessive (or dominant (2.11a)) allele in the new generation of women is equal to a half-sum of these allele frequencies in the previous generation of women and men. Actually, it means that in the association of two female and male X-chromosomes in homozygotes XX, and then formation of a new generation gametes the frequency of the recessive (or dominant) allele will be equal to half of the general frequency of the allele in men and women.

In the process of the crossing-over the frequency of the examined allele in female X-chromosomes has decreased in one since the part of alleles has passed to male X-chromosomes. In male X-chromosomes, on the contrary, the frequency of alleles has increased in one since as a result of the crossing-over part of the alleles of the male X-chromosomes are received from female. Hence, the frequency of allele in the following generation of women according to (2.11) will equal:

$$q_{f(n+1)} = \frac{1}{2}[(1-\beta)q_{fn} + (1+\alpha)q_{mn}] = \\ = \frac{1}{2}[q_{fn} + q_{mn}] - \frac{1}{2}[\beta q_{fn} - \alpha q_{mn}] \qquad (2.21)$$

However, to solve this equation it is obviously not possible in connection with those sizes and the random variables are determined by the crossing-over, i.e., in stochastic they apparently depend on an individual and number of the generation n.

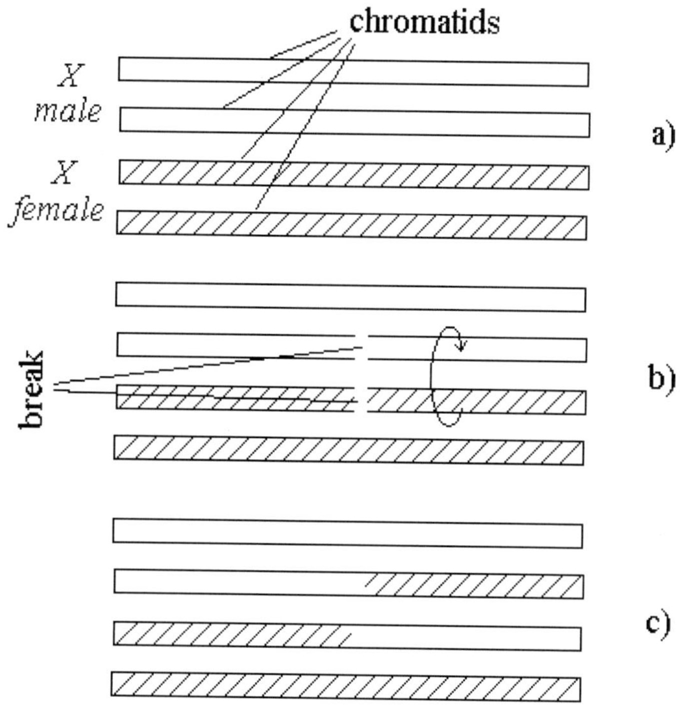

Figure 2.4.

The crossing-over, probably, is not the direct process in a family tree or a population. The crossing-over, apparently, is achieved by the stochastic mixes of the genes in individuals in sex chromosomes doing rather small perturbation of the sizes of allele frequencies in the Hardy-Weinberg law.

There is, at least, one exception. According to researches of the blood groups in men and women in the populations the frequencies of genotypes in these remain constant from generation to generation. Taking into account that blood alleles are in the X-chromosome, it is necessary to assume that crossing-over does not concern these alleles. Thus, there should be sites in the X-chromosomes not subject to crossing-over.

As a whole, the crossing-over constantly restores distinction in the initial allele frequencies of men - for example, q_{m0} and women q_{f0} (2.20) – which leads to the necessity of using the formula (2.11) or is more exact (2.21) instead of (2.4) for each generation.

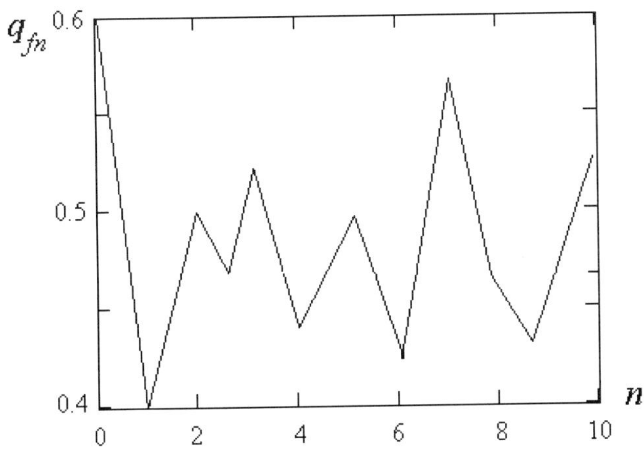

Figure 2.5.

In Figure 2.5 the possible change of the recessive allele *a* frequency q_{fn} in the *X*-chromosome in the woman is shown at the presence of crossing-over. As opposed to the graph in Figure 2.3 the absence of aspiration of this frequency to equilibrium size is observed. The character of the frequency change is rather chaotic.

The purpose of crossing-over there is a sharp increase in the biological variety of one individual sex necessary for the successful evolution of a population.

It is difficult to assume that an autosome's chromatids and *X*-chromosome's chromatids have different properties due to the fact that only crossing-over *X*-chromosomes are subject [13]. It is more probable that in connection with the significance of crossing-over for the evolution in the *X*-chromosomes there is a gene constantly starting the crossing-over.

Let us note that the analysis of the crossing-over in a population on the basis of the stochastic functions theory results in additional conclusions (see paragraph 3.7).

2.2. MUTAGENESIS IN A FAMILY TREE

An extremely important topic of mathematical genetics is the topic of mutations (Latin mutatio-*change*). Mutations explain genetically dependent hereditary diseases. Mutations allow us to explain some processes in the evolution of living organisms. Mutations underlie the selection of animals and plants.

According to a place of the occurrence of the mutations they are classified as [1, 3]:

- *Genomic* mutations resulting in a change in the number of chromosomes;
- *Chromosomal* mutations at which the structure of chromosomes is broken;
- *Genic* mutations which do not result in a change in the number of chromosomes but in the breaking of the normal nucleotides sequence in the DNA molecule. In this book we are basically investigating the genic mutations.

The types of mutation occurring are *induced* and *spontaneous.*

During induced mutagenesis there is an interaction *of the mutagen factor* and an individual which has exposed mutation.

Spontaneous mutations arise spontaneously in the absence of the mutagen factor. They are caused by the existence of the probability of mistakes in the reproduction of a genetic material.

The process of the mutation has a stochastic character. The individuals under action of the mutagen factor with some probability can be subjected to mutations, and others cannot be subjected.

During induced mutation the mutagen factor has the important role. It is possible to classify mutagen factors into two groups: *determined* and *stochastic.*

The determined mutagen factors can be constant or functionally time-dependent.

There is the need to present the following mechanical analogy of the mutation process.

Let us imagine that in a room there is an inflatable ball - representing the mutagen factor. Molecules of air in the room play a role in the individuals. The model mutation is the impact of the molecule with the ball.

The probability of the mutation depends on a state of the molecule; in this case, the distance from the ball. The molecule distant located from a ball can impact it, and the near molecule can depart from the ball during thermal movement not having hit on it. But, the probability of mutation depends on the behavior of the ball. The ball cannot change the volume - the constant mutagen factor.

It can increase at the inflation and reduce at the deflation of the volume. The probability of mutations changes. For example, at complete deflation the ball's probability of mutations becomes equal to zero. The ball can periodically or aperiodically change the volume. All these processes with the ball are models of the determined mutagen factors.

The ball can also change the volume; in completely stochastic it is the model of the stochastic mutagen factor.

We shall consider the variants of genetic inheritance in the action of mutagen factors of various types. Let us start with the analysis of genic mutations of the blood group system ABO in a family tree.

2.3. Spontaneous Mutagenesis in a Family Tree using the Example of Hemophilia

The occurrence of hemophilia is usually attributed to *spontaneous* mutations. However, as this means that a constancy of mutations occur in a family tree (as opposed to a disease due to selection, which would quickly disappear) at the analysis it is possible to examine the problem of hemophilia, assuming the action of some equivalent determined constant mutagen factor.

We research the action of the constant mutagen factor on two-alleles genetically linked with the *X*-chromosome using the example of hemophilia.

Hemophilia is a hereditary disease caused by a deficient or inactive clotting factor of the blood plasma (VIII – hemophilia A, IX –hemophilia B, XI – hemophilia C) necessary for its coagulating. It results in increased fatal bleeding.

The basic proof of the *X*-linked recessive inheritance in hemophilia consists in the fact that in this disease of the blood the sons are damaged, and daughters are phenotypically healthy [3]. There is a connection in the fact that the damaged gene which occurs in the *X*-chromosome in a daughter's zygote is duplicated by the gene of the second homologous *X*-chromosomes where damage is absent (see paragraph 2.1.1, Table 1). In a zygote from which the male individual develops, such duplication is not present. The hemophilia concerns the type of *genic* mutations.

The distribution of genotypes (2.9) which is correct for two-alleles can also be used for the description of blood system ABO (see paragraph 2.1.3). In spite of the fact that this system corresponds to a three-alleles ensemble of the genes, the two alleles *A* and *B* are dominant and their general frequency can be designated in men as p_m, and in women as p_f. Recessive allele O has in this case a frequency in men of q_m, and in women of q_f. The ratio (2.9) for blood system ABO is not the frequency distribution of blood genotypes but the genotype frequency *aa* (or a genotype OO), and also the phenotype frequency corresponding to blood group I truly reflects the ratio.

For the first time the mathematical genetics law has applied Haldane [4] to a problem of hemophilia on the basis of Danforth's idea about the equilibration of the frequency of mutations and selection.

To find the frequency of the recessive allele *a* (or O) in the new generation of a family tree it is necessary to sum the frequency of the occurrence of homozygotes *aa* and half of the frequency of the occurrence of heterozygotes *Aa* from the ratio (2.9) in the previous generation (see (2.11)). Using this method of calculation of the allele frequencies we find the frequency of the variation of allele *a* in female generations of carriers of a hemophilia gene equals:

$$q_{fn} = \frac{1}{2}[q_{f(n-1)} + q_{m(n-1)}] - b, \tag{2.22}$$

where *b* is some constant which we shall name the equivalent constant mutagen factor. This factor specifies that the frequency of healthy allele *a* in a woman of the subsequent generation for hemophilia to become less than under the absence of the condition of mutagenesis (2.11). The frequency of healthy allele *a* in a female part of a family tree for hemophilia during the lifespan of a family tree is reduced due to the reduction of biological fitness first of all in the male individuals. This results in a decrease in the general reproductive ability in a family tree [3].

We use a condition (2.12) whereby the frequency of alleles in the *X*-chromosome of the woman is equal to the frequency of the same alleles in the man in the following generation:

$$q_{m(m-1)} = q_{f(n-2)}. \tag{2.23}$$

Substituting (2.23) in (2.22), we find:

$$q_{fn} = \frac{1}{2}\left(q_{f(n-1)} + q_{f(n-2)}\right) - b. \quad (2.24)$$

To solve the recurrent ratio (2.24) we shall write it down as:

$$q_{f(n-1)} - 2q_{fn} + q_{f(n-2)} = 2b. \quad (2.25)$$

The solution to the finite-differential equations (2.25) we find as $q_{fn} = a^n - \frac{2}{3}bn$ (actually as the sum of general and particular $q_{fn} = -\frac{2}{3}bn$ solutions) where in this case a is constant. Substituting this solution in the formula (2.25), we exclude size b:

$$a^{n-1} - \frac{2}{3}b(n-1) - 2\left(a^n - \frac{2}{3}bn\right) + a^{n-2} - \frac{2}{3}b(n-2) = 2b. \quad (2.26)$$

Collecting similar members, we find:

$$a^{n-1} - 2a^n + a^{n-2} = 0.$$

Let us divide the equation by a^{n-2}:

$$a - 2a^2 + 1 = 0. \quad (2.27)$$

We find two roots of the characteristic quadratic (2.27):

$$a_1 = 1 \text{ and } a_2 = -\frac{1}{2}. \quad (2.28)$$

Writing down the general solution as $q_{fn} = C_1 a_1^n + C_2 a_2^n$, we find the solution to the finite-differential equations (2.25):

$$q_{fn} = C_1 + C_2\left(-\frac{1}{2}\right)^n - \frac{2}{3}bn. \quad (2.29)$$

We shall find constants of integration C_1 and C_2 on the basis of the initial conditions: at $n = 0$, $q_{fn} = q_{f0}$ and at $n = 1$ according to (2.22) $q_{fn} = q_{f1} = \frac{q_{m0} + q_{f0}}{2} - b$. Thus:

$$C_1 = \frac{2q_{f0} + q_{m0}}{3} - \frac{2}{9}b \text{ and } C_2 = \frac{q_{f0} - q_{m0}}{3} + \frac{2}{9}b.$$

Therefore, the solution (2.29) finally looks like:

$$q_{fn} = \frac{2q_{f0} + q_{m0}}{3} + \left(\frac{q_{f0} - q_{m0}}{3} + \frac{2}{9}b\right)\left(-\frac{1}{2}\right)^n - \frac{2}{3}b\left(n + \frac{1}{3}\right) \quad (2.30)$$

In Figure 2.6 the graph of the allele *a* (or O) frequency dependence q_{fn} in generations of women is shown. For the generality of calculation the allele frequencies in women and men are accepted as being initially unequal. In calculation they have also been accepted for Figure 2.3: $q_{f0} = 0.6$ and $q_{m0} = 0.2$. The equivalent constant mutagen factor of hemophilia is accepted as $b = 1.95 \cdot 10^{-4}$. The chosen value of the mutagen factor will be proved further in paragraph 3.3 at the analysis of a population as a whole.

Apparently, from the graph the frequency of allele *a* (or O) in the *X*-chromosome of the woman gradually falls. For comparison, the equilibrium frequency of allele *a* at $b = 0$ and $n \to \infty$ is shown by the dotted line. It is equal according to (2.20) or (2.30) size $q_{f\text{рав}} = \frac{2q_{f0} + q_{m0}}{3} \approx 0.467$.

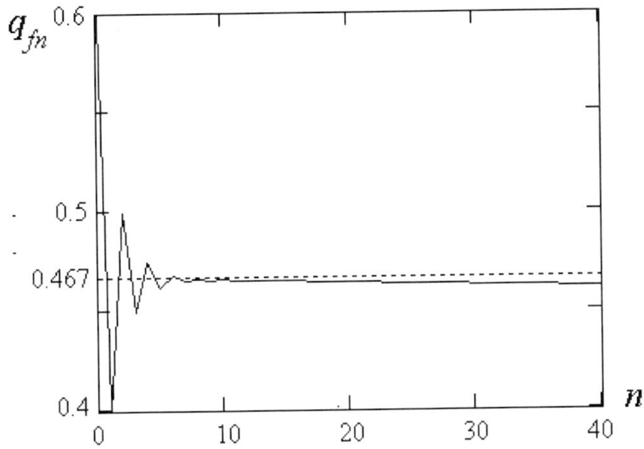

Figure 2.6.

Actually, as was specified earlier, see paragraph 2.1.5, locus in which are located genes of blood in *X*-chromosomes not subject to a crossing-over. Therefore, it is necessary to believe that the allele frequencies in men and women in generations are identical $q_{fn} = q_{mn} = q_n$. Formally, inheritance of the blood alleles is carried out similarly to autosomal inheritance (2.4). The recurrent equation for such an inheritance in view of the equivalent constant mutagen factor b_1 looks like:

$$q_n = q_{n-1} - b_1. \tag{2.31}$$

We find the solution to this equation by the method of consecutive substitution:

$$q_n = q_0 - b_1 n. \tag{2.32}$$

In Figure 2.7 the graph plotted under the formula (2.32) at initial size of the blood allele O frequency equal to $q_0 = 0.605$ (see (2.8)) is shown.

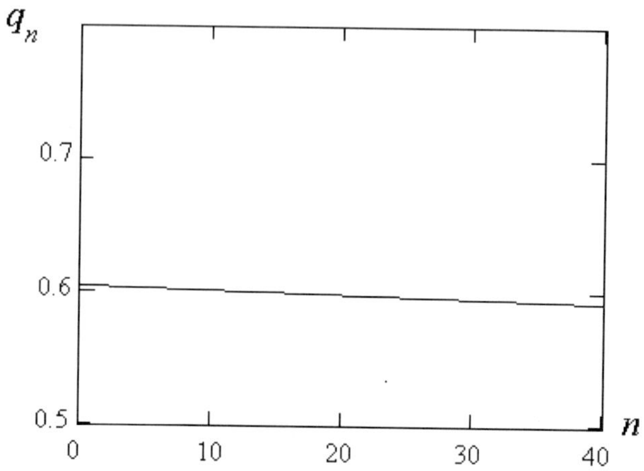

Figure 2.7.

As opposed to Figure 2.6, in Figure 2.7 there are no initial oscillations of frequency q_n of examined recessive allele O.

There is a uniform decrease in the frequency of the healthy blood alleles in a family tree.

In addition, the determination of the equivalent constant mutagen factor in formulas (2.22) and (2.31) differ from each other a little. The mutagen factor determined according to the formula (2.22) is 1.5 times greater than that determined under the formula (2.31). At $n \to \infty$ and equality of the male and female alleles the initial frequencies $q_{f0} = q_{m0}$ from comparison (2.30) and (2.32) follows $\frac{2}{3}b = b_1$. Figure 2.7 is plotted for value $b_1 = 1.3 \cdot 10^{-4}$.

Variation of the allele blood O frequency (shown in Figure 2.6) can arise by association in one space of two populations or family trees in which equilibrium allele frequencies in individuals initially are not identical. However, through some generations $n \approx 10$ arises a new population community (on blood alleles) and variation of the allele blood O frequency starts to occur in conformity (Figure 2.7).

The chosen value of the equivalent constant mutagen factor is an average for the population. However, in various family trees this factor can differ significantly between each other and the size of this factor can be much greater.

In Figure 2.8, according to [3, 24], the family tree of the European royal houses is shown, in which men frequently suffered from hemophilia. The distribution of the *X*-chromosome in a family tree is shown also. In this family tree of Queen Victoria (who was born in 1819) mutant allele O has arisen *de novo*. In the figure the circle leads round the *X*-chromosome having destructive alleles O, which causes hemophilia. The black squares there indicate men who suffered from hemophilia. From this family tree the gradual increase in the number of men with hemophilia is clearly visible. The family tree shown is not complete, since it is possible to conclude that in the third generation all of the men were suffering from hemophilia. Actually, that was far from being the case. And now in the royal dynasty in question the proportion of hemophiliac men is not too great.

The formula (2.22) can be written down as:

$$q_{fn} = \frac{1}{2}\left[\left(q_{f(n-1)} - b\right) + \left(q_{m(n-1)} - b\right)\right]. \qquad (2.33)$$

Hence, the equivalent mutagen factor determines a reduction of the proportion of healthy men (or women) in a family tree for one generation.

According to the selected family tree, in the first generation one of the four men suffers from hemophilia, and in the second generation three men out of seven are hemophiliacs. A rough calculation of the increase in the frequency of hemophilia in males in the dynasty in question in the first and second generations shows $\frac{3}{7} - \frac{1}{4} = 0.178$. The proportion of healthy men has decreased by the same amount in the first and second generations also.

Therefore, for the family tree shown in Figure 2.8 it is initially possible to use the equivalent mutagen factor of hemophilia $b = 0.178$.

In Figure 2.9 the calculation under the formula (2.30) variation of frequency q_{fn} of the intact allele *a* (or O) in generations of women of the European royal houses' family tree is shown. As follows from the calculation, after 4 - 5 generations in the family tree the women do not remain genotypically healthy.

It is necessary to note that the performed calculation has a very conditional character. First, selection is not taken into account. Second, the intensive receipt of healthy men and women from other family trees is poorly taken into account. Thirdly, it is not taken into account that men who are suffering from hemophilia can give posterity, and cannot participate in this process.

We shall postpone more detailed analysis and calculation of the equivalent mutagen factor genetic sense of a hemophiliac to paragraph 3.3, where its size is found to be $b = 0.214$. For this size *b* the calculation of frequency q_{fn} variation is also shown in Figure 2.9.

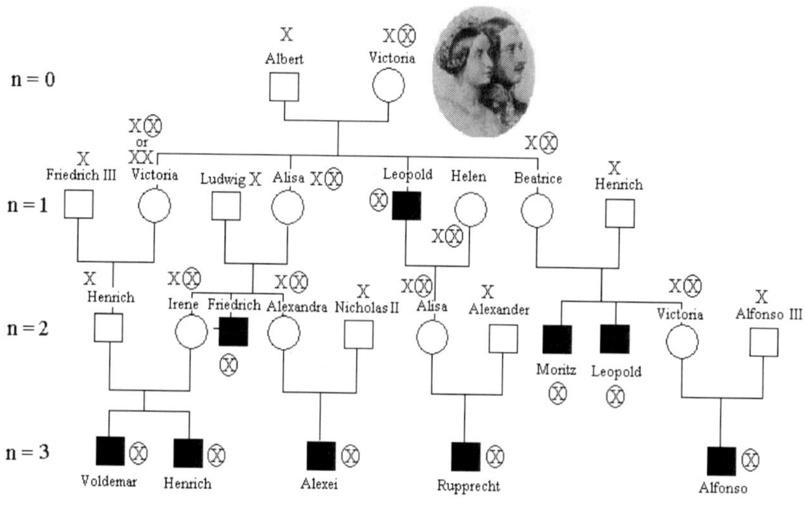

Figure 2.8.

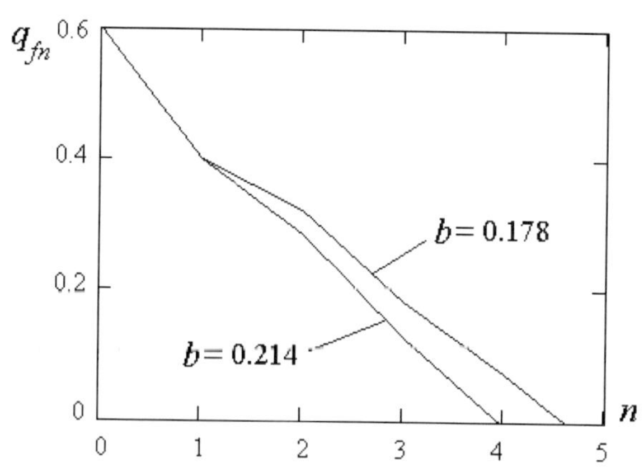

Figure 2.9.

Thus, in the presence of the equivalent constant mutagen factor the equilibrium frequency of allele a (or O) is not established. It is connected to absence in the calculation of selection which will be considered in the following paragraph.

2.4. Natural Selection in a Family Tree

In a family tree where hemophilia is observed, the natural selection operates, resulting in a decrease of genetic frequencies; in particular, the frequencies of damaged allele O in one direction.

Natural selection is expressed in the primary survival and retention of posterities by the most fitted individuals (genotypes).

As opposed to natural selection, artificial selection is organized by the human purposefully for reception of the most economically valuable animals or plants. We shall not analyze this kind of selection.

The action of selection is intensive enough. For example, the life span of men with hemophilia is approximately 1/3 of the life span of healthy people [3], so male fertility (i.e., the chances of having posterity) in comparison with healthy men is reduced. Therefore, not all men of the given family tree participate in the reception of posterity and damaged alleles O are eliminated from the family tree. However, the balance of mutagenesis and selection resulting in the preservation on some level of the Hardy-Weinberg equilibrium or ratio of the blood genotype frequencies (2.5) is possible.

The balance of the mutagenesis and selection in hemophilia allows us to find an interrelationship between the equivalent constant mutagen factor b and standard parameter of selection s [3].

Let us consider selection against homozygotes aa (or OO). The results of the given analysis can be applied to hemophilia. Similarly, selection as opposed to other types of zygotes is calculated.

Genotypes before selection, for example, in generation $n-2$, are distributed according to (2.9).

The *fitness* of a genotype characterizes the ability (probability) of the given genotype to reproduce.

We accept the ratio of fitnesses of genotypes AA, Aa, aa [3]:

$$1:1:(1-s). \tag{2.34}$$

where s is the reduction of the recessive homozygotes OO fraction as a result of the reduction in individual reproductive ability, the so-called standard parameter of selection. The size $1-s$ refers to the fitness of the given genotype.

Genotypes after selection we shall write down for the following generation $(n-1)$:

$$(AA)\, p_{f(n-1)}p_{m(n-1)} : (Aa)\left(p_{m(n-1)}q_{f(n-1)} + p_{f(n-1)}q_{m(n-1)}\right) : \\ : (aa)\, q_{m(n-1)}q_{f(n-1)}(1-s) \quad (2.35)$$

Taking into account that

$$p_{f(n-1)}p_{m(n-1)} + \left(p_{m(n-1)}q_{f(n-1)} + p_{f(n-1)}q_{m(n-1)}\right) + q_{m(n-1)}q_{f(n-1)} = \\ = \left(p_{f(n-1)} + q_{f(n-1)}\right)\left(p_{m(n-1)} + q_{m(n-1)}\right) = 1$$

we find the sum of genotype frequencies: $1 - q_{m(n-1)}q_{f(n-1)}s$.

Furthermore, using a standard rule of a finding of allele frequency in the following generation (half of a heterozygotes frequency plus the frequency of homozygotes) and the formula (2.35), we calculate the frequency of the recessive allele a (or O) in women in generation n which is similar [3] when such a calculation is carried out for autosomal genome:

$$q_{fn} = \frac{\frac{1}{2}\left(p_{m(n-1)}q_{f(n-1)} + p_{f(n-1)}q_{m(n-1)}\right) + q_{m(n-1)}q_{f(n-1)}(1-s)}{1 - q_{m(n-1)}q_{f(n-1)}s} \quad (2.36)$$

In (2.36) normalization of the general frequency of all genotypes $1 - q_{m(n-1)}q_{f(n-1)}s$ is carried out.

Let us transform the formula (2.36) using $p_{m(n-1)} = 1 - q_{m(n-1)}$ and $p_{f(n-1)} = 1 - q_{f(n-1)}$:

$$q_{fn} = \frac{\frac{1}{2}\left(q_{f(n-1)} + q_{m(n-1)}\right) - q_{m(n-1)}q_{f(n-1)}s}{1 - q_{m(n-1)}q_{f(n-1)}s}. \tag{2.37}$$

For the finding of an interrelationship between the equivalent constant mutagen factor b and standard parameter of selection s we use Danforth's idea [5] about an equilibration of the mutations' frequency and selection. Thereof we shall equate (2.37) and (2.22):

$$\frac{\frac{1}{2}\left(q_{f(n-1)} + q_{m(n-1)}\right) - q_{m(n-1)}q_{f(n-1)}s}{1 - q_{m(n-1)}q_{f(n-1)}s} = \frac{1}{2}[q_{f(n-1)} + q_{m(n-1)}] - b \tag{2.38}$$

After simple transformations with the account (2.22) we find:

$$s = \frac{b}{q_{m(n-1)}q_{f(n-1)}\left(1 - q_{fn}\right)}. \tag{2.39}$$

Let us estimate the standard parameter of selection for a family tree, shown in Figure 2.8. Supposing that generations $n = 2$ and $n-1=1$, we shall find $q_{m(n-1)}$. The total of men, and hence X-chromosomes, in the first generation equals 4. Healthy men equal 3. Hence, $q_{m(n-1)} = \frac{3}{4}$. The total of X-chromosomes in women in the first generation equals 8. The total number of healthy X-chromosomes equals 5 or 4, since the situation with Victoria is not clear. Hence, $q_{f(n-1)} = \frac{5}{8}$ or $q_{f(n-1)} = \frac{1}{2}$. In the second generation obviously frequency healthy alleles a (or O) in women equal $q_{fn} = \frac{1}{2}$. Substituting the received frequencies of X-chromosomes (or alleles a) in the formula (2.39) under condition of $b = 0.178$, we shall find standard parameter of selection $s = 0.76 - 0.95$. Hence, fitness of the blood genotype OO in a family tree, Figure 2.8, equals $1 - s = 0.24 - 0.05$.

The estimation of fitness is enough challenge. One of the possible ways to calculate fitness is to find it as the ratio of the survived children as an average of the falling one parent (male) who suffers from hemophilia to the average of survived children falling on one healthy parent (male). In [6] the value of fitness in hemophilia is given as a result of the statistical analysis of two generations (cohorts [6]) in a population. Using the data on the number of survived children from 100 healthy men – reproducers - 240 and on 100 hemophiliac men - reproducers – 70 the value of fitness $1-s=\frac{70}{240}=0.29$ is received.

Let us note the lack of the actual mixture at the calculation of fitness offered in [6] estimations of the fitness in the family trees where hemophilia is observed. Thus, it is not taken into account that the fitness of individuals in various family trees may differ. More correctly, we should speak about the fitness of individuals of a particular family tree.

2.5. Induced Mutagenesis in a Family Tree as an Example of Radiating Influence

Disasters of nuclear installations (Chernobyl, Fukushima) and the results of the possible application of nuclear weapons lead to a raised radioactive background being a source of the mutations affecting a genofund of a human population. Therefore, it is of interest to consider the existence and development of the population in conditions influenced by the mutagen ionizing radiation which gradually decreases according to the law of radioactive decay of the elements generating radioactive radiation. Thus, action on the population of the determined mutagen factor exponentially decreasing over time is examined.

Failures of the genetic apparatus have arisen which are generally incompatible with the normal further functioning of the organism. Mutations in genes in sex chromosomes can lead either to the disappearance of reproductive function in individuals or to the appearance of serious failure in the offspring.

We will analyze the existence of the population in the radiation environment, and mathematical modeling of changes in the genofund, taking as an example changes in both sex-linked allele frequencies and those inherited autosomally.

2.5.1. Action of Radiation on the Autosomal Genome

Let's examine the two-allele genetic system. Let us imagine that in some generation there is a ratio of dominant and recessive alleles:

$$(A)\,p:(a)\,q, \tag{2.40}$$

where the frequency of dominant allele A equals p, and the frequency of recessive allele a equals q.

It is natural to assume that when individuals of different sexes mate the radioactive influence is carried out on all genotypes of the autosomal genome at association of gametes in zygotes: on dominant AA and recessive aa homozygotes, and also on heterozygotes Aa. Thus, we shall assume that change of dominant homozygotes AA frequency p^2+Fpq in the following generation equals α, of heterozygotes Aa frequencies $2pq(1-F)$ equals β, and of recessive homozygous aa frequencies q^2+Fpq equals γ. We can assume that during the impact of radiation on the family tree the frequencies of genotypes in the following generation have changed in proportions:

$$\begin{aligned}&(AA)(p^2+Fpq+\alpha):(Aa)[2pq(1-F)+\beta]:\\&:(aa)(q^2+Fpq+\gamma)\end{aligned}, \tag{2.41}$$

that breaks the Hardy-Weinberg relation (2.2) (or taking into account inbreeding – Wright [6]). In ratio (2.41) the condition of panmixia, see paragraph 1, is weakened. There is an opportunity for consanguineous marriages with inbreeding coefficient F [7]. This factor characterizes a reduction in individuals of the inbred populations - see paragraph 1 (more exact, there is a family tree) - and shares of heterozygotes in $(1-F)$ in comparison with a panmictic population. In more detail, the phenomenon of inbreeding and its consequences will be investigated in paragraph 2.7.

In the next generation the Hardy-Weinberg equilibrium is restored with the new ratio of allele frequencies. To obtain the dominant frequency of allele A in the new balance we need to sum up the frequency of homozygotes AA and halve the frequency of heterozygotes Aa from (2.41). Similarly, we will do this for the allele a. As a result, we obtain:

$$(A)(p+\alpha+\beta/2):(a)(q+\gamma+\beta/2). \quad (2.42)$$

Thus, it is natural that the condition $p+\alpha+\beta/2+q+\gamma+\beta/2=1$ is observed in each generation.

Thus, the ratio of the frequencies of genotypes in the next generation is:

$$(AA)(p+\alpha+\beta/2)^2:(Aa)2(p+\alpha+\beta/2)(q+\gamma+\beta/2):$$
$$:(aa)(q+\gamma+\beta/2)^2 \quad (2.43)$$

Let us note that in (2.42) and (2.43) there is not an inbreeding coefficient *F*.

In the absence of influences on the family tree, i.e., $\alpha=\beta=\gamma=0$, the ratio (2.43) is identical to (2.2), i.e., we have a well-known position to maintain the frequency of genotypes in a set of generations.

The question of the long gradual reduction of the radiation exposure in which the family tree evolves is of great interest.

Let us consider a model situation of a family tree in a region where there is only one radioactive element with the half-life $T_{1/2}$. It was established [3] that the frequency of mutations (usually lethal) as a first approximation is proportional to the power dose of radiation $\mu=k_1P$, where k_1 is a constant. This law is correct at rather small mutational frequencies and doses of radiation.

As the power dose of radiation is proportional to the activity of radioactive elements scattered in the environment, it is possible to record the frequency μ of mutations using the fundamental of radioactive decay law as the ratio $\mu\sim\left(\frac{1}{2}\right)^{\left(\frac{T}{T_{1/2}}\right)n}$ where $T\sim25$ years is the approximate duration of one generation alternation, *n* is the number of the examined generation existing in the conditions of radiation.

As the law of action on the population has a functional character, the mutagen influence is determined; i.e., the determined mutagen factor operates on the family tree.

Accepting $R = \left(\frac{1}{2}\right)^{\left(\frac{T}{T_{1/2}}\right)}$ and considering the influence of radiation on recessive homozygotes *aa* only, i.e., $\mu = \gamma + \beta/2$, we obtain the mathematical model of mutational process:

$$\gamma + \beta/2 = (k/2)R^n, \tag{2.44}$$

where the coefficient 1/2 is taken for the convenience of further transformations. The coefficient k characterizes the initial activity of the radioactive elements that affect the family tree. In accordance with [8] $k \sim -\frac{N_0}{T_{1/2}}$, where N_0 is the total initial quantity of the radioactive substance that acts on the family tree.

For example, there are recessive genes responsible for the content of catalase in the blood. One can cite as an example of reducing the frequency of recessive alleles of these genes at 15.5 ones for about two generations of people living in Hiroshima and Nagasaki in comparison with the general population of Japan. In this case, the frequency of dominant alleles has not changed [7].

Based on (2.42) and (2.44), the recursive formula for changing the frequency of allele *a* for the generations of the family tree is as follows:

$$q_n = q_{n-1} + \gamma + \beta/2 = q_{n-1} + (k/2)R^n, \tag{2.45}$$

where $k < 0$ is the frequency of healthy allele falls, $R < 1$.

The solution of this equation is not difficult and it is the sum of a geometric progression with common quotient R:

$$q_n = q_0 + \frac{k}{2}\frac{R(1-R^n)}{1-R}, \tag{2.46}$$

where q_0 is the frequency of allele *a* in the initial generation.

If the number of generations tends to infinity $n \to \infty$, the frequency of allele *a* tends to the value:

$$\hat{q} = q_0 + \frac{k}{2}\left(\frac{R}{1-R}\right). \tag{2.47}$$

The possibility of establishing equilibrium is determined by the condition $q_0 > \left|\frac{k}{2}\frac{R}{(1-R)}\right|$. Taking $q_0 = 0$, we can solve equation (2.46) relatively to *n* a number of generations for which will complete the elimination of the allele *a*. It reduces the variability of the family tree's genofund.

In Figure 2.10 there is a graph of the dependence frequency q_n of allele *a* on the number of generations *n* plotted in accordance with the formula (2.46) for the half-life of radioactive elements in the region accordingly: curve 1 - $T_{1/2} = 176$; curve 2 – 93.8 and curve 3 - 22.6 years. For example, the half-life of widespread radioactive substances is such as Cs_{55}^{137} – 30.2 years, Sr_{38}^{90} – 28.7 years, and Co_{27}^{60} - 5.3 years [9]. But usually a half-life isotope is considerably smaller.

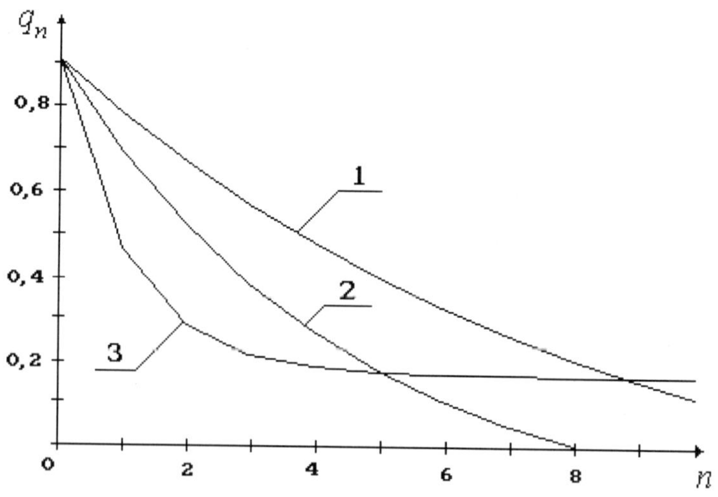

Figure 2.10.

As can be seen from the graphs for the half-life 93.8 years, the allele a will disappear in the eighth generation. For the half-life $T_{1/2} = 22.6$ years after a fifth-generation set a new equilibrium frequency of allele a $\hat{q} = 0.17$. In the calculations for allele a the initial frequency $q_0 = 0.9$ and constant $k = -\frac{50}{T_{1/2}}$ are used.

There is a critical size of the half-life $T_{1/2} = 37.4$ years above which the Hardy-Weinberg equilibrium is not established as a result of complete elimination of the allele a.

Figure 2.10 shows that for a small half-life, curve 3, the frequency of alleles a first drops very quickly because of the high activity radioactive elements [25]. But then equilibrium is established because the rapid decrease in the quantity of highly active substances also reduces the effects of radiation. For the long half-life the frequency of allele a decreases more slowly, but the effect of long-term radiation exposure completely eliminates the allele (curves 1 and 2).

2.5.2. Action of Radiation on Genome Linked to the *X*-chromosome

Let us consider the influence of radioactive radiation on a gene linked to the *X*-chromosome. Let us suppose that A and a alleles are linked to the X chromosome. The distribution of genotypes in the absence of the mutagen factor's influence corresponds in women (2.9). We will assume as well that in the case of autosomal inheritance on the family tree the external mutagen factor changing the ratio of frequencies of genotypes in women is operating:

$$\begin{aligned}
&(AA)[p_f p_m + (p_m q_f + p_f q_m)F/2 + \alpha]: \\
&:(Aa)[(p_m q_f + p_f q_m)(1-F) + \beta]: \\
&:(aa)[q_m q_f + (p_m q_f + p_f q_m)F/2 + \gamma],
\end{aligned} \tag{2.48}$$

where F is the inbreeding factor.

Then, using a method of the allele frequencies calculation in the following generation consisting in addition to the half of heterozygotes *Aa* frequency and homozygotes *aa* frequency in (2.48), we find the frequency of allele *a* in generations of women equally:

$$q_{fn} = \frac{1}{2}[q_{f(n-1)} + q_{m(n-1)}] + \gamma + \beta / 2. \tag{2.49}$$

As in the case of autosomal inheritance the inbreeding coefficient F in relation (2.49) is not included.

As in the case of autosomal inheritance we examine the effects of ionizing radiation on the recessive homozygotes *aa*.

According to (2.44), the equation (2.49) becomes:

$$q_{fn} = \frac{1}{2}[q_{f(n-1)} + q_{f(n-2)}] + (k/2)R^n. \tag{2.50}$$

At record (2.50) it is taken into account that the man inherits the *X*-chromosome from his mother, i.e., $q_{m(n-1)} = q_{f(n-2)}$.

The solution to this equation is the sum of the general solution to the homogeneous equation (2.51) and a particular solution to the inhomogeneous equation (2.50), [10]:

$$\tilde{q}_{fn} = \frac{1}{2}[\tilde{q}_{f(n-1)} + \tilde{q}_{f(n-2)}], \tag{2.51}$$

so:

$$q_{fn} = \tilde{q}_{fn} + q^*_{fn}. \tag{2.52}$$

The general solution to the homogeneous equation (2.51) $\tilde{q}_{fn}$ looks similar (2.18):

$$\tilde{q}_{fn} = C_1 + C_2\left(-\frac{1}{2}\right)^n, \tag{2.53}$$

where C_1 and C_2 are constants. Solution (2.53) with the initial conditions describes the change of the allele *a* frequency in women in the generations in the absence of external mutagen factors on the family tree (see paragraph 2.1.4).

A particular solution to the equation (2.50) will be sought in the form $q^*_{fn} = LR^n$ where *L* is yet unknown. Substituting this formula into (2.50), we find:

$$2LR^n - LR^{n-1} - LR^{n-2} = kR^n. \tag{2.54}$$

Dividing both sides by R^{n-2} we find the value of *L*:

$$L = \frac{kR^2}{2(R-1)(R+1/2)}. \tag{2.55}$$

Thus, the solution to equation (2.50) in accordance with (2.52) is function:

$$q_{fn} = C_1 + C_2\left(-\frac{1}{2}\right)^n + \frac{kR^{n+2}}{2(R-1)(R+1/2)}. \tag{2.56}$$

Constants C_1 and C_2 can be found on the basis of the initial conditions: at $n = 0$ the frequency $q_{fn} = q_{f0}$ and at $n = 1$ according to (2.49) and (2.44),

$$q_{f1} = (q_{m0} + q_{f0})/2 + \gamma + \beta/2 = (q_{m0} + q_{f0})/2 + (k/2)R.$$

Substituting initial conditions in (2.56), we shall find:

$$C_1 = \frac{2q_{f0} + q_{m0}}{3} - \frac{kR}{3(R-1)}$$

and

$$C_2 = \frac{q_{f0} - q_{m0}}{3} - \frac{kR}{6(R+1/2)}. \quad (2.57)$$

Consequently, the solution to equation (2.50) is:

$$q_{fn} = \frac{2q_{f0} + q_{m0}}{3} - \frac{kR}{3(R-1)} + \left[\frac{q_{f0} - q_{m0}}{3} - \frac{kR}{6(R+1/2)}\right]\left(-\frac{1}{2}\right)^n + \quad (2.58)$$
$$+ \frac{kR^{n+2}}{2(R-1)(R+1/2)}$$

If the number of generations tends to infinity ($n \to \infty$), there is an equilibrium frequency of allele *a* in women (if it is possible):

$$\hat{q}_{fn} = \frac{2q_{f0} + q_{m0}}{3} - \frac{kR}{3(R-1)}. \quad (2.59)$$

Similarly to the case of autosomal inheritance at $q_{fn} = 0$, the number of generations which would completely eliminate allele *a* in women can be found.

Figure 2.11 shows a graph of the dependence of the allele *a* frequency q_{fn} in women and the number of generations *n* plotted by the formula (2.58) for the same half-life as for autosomal inheritance (see paragraph 2.5.1).

As can be seen from the graphs, for the half-life $T_{1/2} = 93.8$ years - the allele *a* will disappear in the ninth generation (curve 2). For the half-life $T_{1/2} = 22.6$ years since the sixth generation the new equilibrium is established with the frequency of allele *a* in women $\hat{q}_{fn} = 0.14$, curve 3. Curve 1 is designed for the half-life of $T_{1/2} = 176$ years. At the calculations used, the initial frequency of allele *a* in women $q_{f0} = 0.9$; in men $q_{m0} = 0.1$.

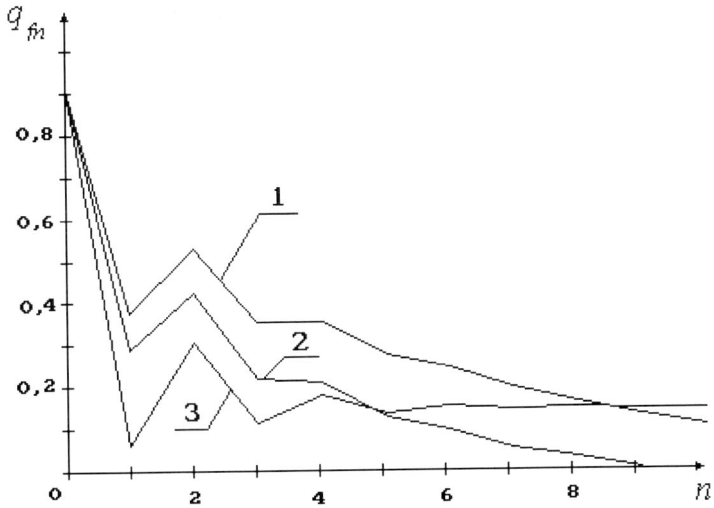

Figure 2.11.

The critical size of the half-life above which the Hardy-Weinberg equilibrium is not established as a result of the complete elimination of the allele a in the case of the genes linked to the X chromosome inheritance there is $T_{1/2} = 44.7$ years.

Figure 2.12 shows graphs of the dependence of the generation number n_k of the allele a complete elimination, and half-life $T_{1/2}$ of radioactive element for autosomal genes - 1 and for the sex-linked genes - 2. The graphs are based on the solutions of equations (2.46) and (2.58) at conditions accordingly $q_n = 0$ and $q_{fn} = 0$. The graphs show that there are more dangerous radioactive elements to the population. For the given model calculation there are the elements with half-life $T_{1/2} = 52.2$ years for autosomal genes and $T_{1/2} = 60.2$ years for genes linked to sex. In the first case the complete elimination of the allele a will occur after $n_k \approx 6$ generations, and in the second case - after $n_k \approx 8$ generations. Thus, sex-linked genes are more resistant to radiation than autosomal genes.

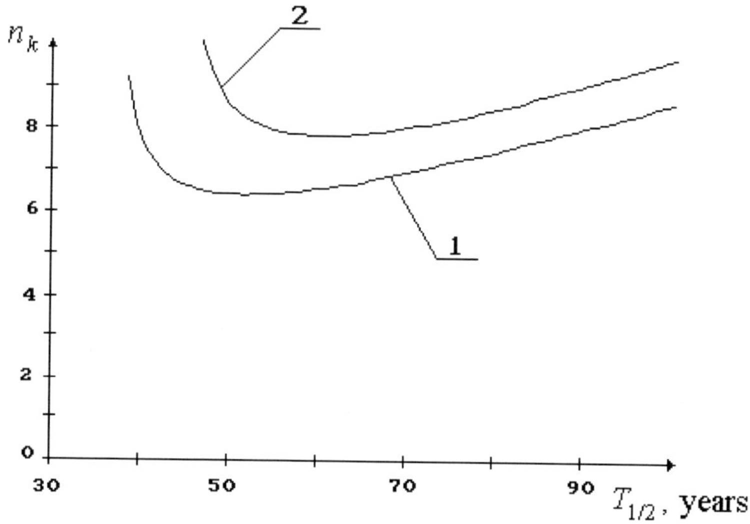

Figure 2.12.

As a result of the lead mathematical modeling of the genetic processes, it is possible to draw the following conclusions:

a) If there is the action of radiation on a family tree, two outcomes are possible depending on the half-life of radioactive elements in the environment:
   - Establishment of the Hardy-Weinberg equilibrium in maintaining the existing genome but at a lower level of the targeted allele frequencies;
   - The disappearance of the targeted allele, and consequently the impossibility of preserving the Hardy-Weinberg equilibrium.

b) There is a boundary size of the half-life of radioactive substances in the environment below in which the family tree maintains its genome. Consequently, the outcome of the family tree depends on the nature of the radiation environment (disaster at nuclear power plants, nuclear explosion, accident at work, natural phenomenon, etc.).

c) Sex-linked genes have a somewhat higher resistance to radiation compared with autosomal genes.

The obtained results allow us to be more cautious when facing the choice of whether to use radioactive elements with specific half-lives which have a strong mutagenic effect on autosomal genes and linked to the X chromosome genes.

Considering that, in evolutionary terms, genes linked to the X-chromosome have greater resistance, we can recall that genes are located in the X-chromosome, the mutations of which affect the vital functions of the human organism (vision, blood clotting, skin and etc. [3]). Therefore, these genes are better protected against the action of mutagens such as radiation. Greater resistance to the radiation of alleles which are localized in 23 pairs of chromosomes again suggests that the female organism is genetically more stable. It plays an important role in maintaining the size of the population.

The obtained results allow us to determine the radiation conditions of maintaining the Hardy-Weinberg balance in the population as well as conditions for the complete elimination of the targeted allele by ionizing radiation.

2.6. Incomplete Inheritance in a Family Tree

In genetics there is an interesting phenomenon called incomplete inheritance. We shall discuss the expediency of the inheritance of such a kind of existence below during the analysis of this phenomenon in a population.

In the given section we shall analyze the mathematical aspects of incomplete inheritance in a family tree. The biological mechanism of such an inheritance can be the following.

The existence of the effect of genome incomplete replication [1] is known. Replication of DNA occurs by fragments which have 1000 - 2000 nucleotides, so-called Okazaki fragments [1].

The information on the maximum quantity of replicated Okazaki fragments can be written down in the genome and the corresponding gene and in some cases the replication of DNA can start to regulate the necessary quantity of fragments. As will be shown below, the number of the replicated Okazaki fragments allows the population to choose a particular kind of process of frequency change of the recessive alleles.

Let us consider the equation (2.11) connecting in generations the recessive allele frequency in the male and female X-chromosome in the following kind:

$$q_{fn} = \frac{1}{2}[q_{f(n-1)} + q_{m(n-1)}]. \qquad (2.60)$$

where $q_{f(n-1)}$ and $q_{m(n-1)}$ are frequencies of the allele in the woman and the man in the previous $n-1$ generation.

Let us assume that the man inherits the *X*-chromosome from the mother incompletely:

$$q_{m(n-1)} = (1-\alpha)q_{f(n-2)}. \tag{2.61}$$

The size α, which we shall name the parameter of the incomplete inheritance, shows the fraction of recessive alleles the man does not inherit from mother.

Substituting (2.61) in (2.60), we receive:

$$q_{fn} = \frac{1}{2}[q_{f(n-1)} + (1-\alpha)q_{f(n-2)}]. \tag{2.62}$$

The solution to the equation (2.62) we shall find as $q_{fn} = a^n$, where *a* in this case is a constant. Substituting this expression in (2.62), we shall find:

$$a^n = \frac{1}{2}[a^{n-1} + (1-\alpha)a^{n-2}]. \tag{2.63}$$

Having divided both parts of the equation on a^{n-2}, we shall receive some characteristic equation:

$$a^2 - \frac{1}{2}a - \frac{1-\alpha}{2} = 0, \tag{2.64}$$

whose solution are two roots $a_1 = \frac{1}{4}(1+\mu)$ and $a_2 = \frac{1}{4}(1-\mu)$ where $\mu = \sqrt{9-8\alpha}$ is a constant also describing incomplete inheritance. The size μ is in a range $1 \le \mu \le 3$ at $1 \ge \alpha \ge 0$. At $\alpha = 0$ or $\mu = 3$ the equation (2.62) describes an equilibrium condition of the genome in the absence of incomplete inheritance or, more exactly, the approach to the Hardy-Weinberg equilibrium. At $\alpha = 1$ or $\mu = 1$ the inheritance on a male line is absent, which apparently is impossible.

The general solution to the equation (2.62) looks like:

$$q_{fn} = C_1 \left(\frac{1}{4}(1+\mu) \right)^n + C_2 \left(\frac{1}{4}(1-\mu) \right)^n, \tag{2.65}$$

where C_1 and C_2 are constants. The solution (2.65) in view of the initial conditions describes the frequency of allele *a* in women in generations at the account of the incomplete inheritance factor.

We shall find constants C_1 and C_2 proceeding from the initial conditions: at $n = 0$, $q_{fn} = q_{f0}$ and at $n = 1$ according to (2.60) $q_{fn} = q_{f1} = \dfrac{q_{m0} + q_{f0}}{2}$. Hence:

$$C_1 = \frac{2q_{m0} + (1+\mu)q_{f0}}{2\mu} \text{ and } C_2 = -\frac{2q_{m0} + (1-\mu)q_{f0}}{2\mu}.$$

Substituting constants C_1 and C_2 in the formula (2.65), we find the solution to the equation (2.62) as:

$$q_{fn} = \frac{1}{2\mu 4^n}\left(2q_{m0}\left((1+\mu)^n - (1-\mu)^n\right) + q_{f0}\left((1+\mu)^{n+1} - (1-\mu)^{n+1}\right)\right) \tag{2.66}$$

where for the further transformations it is convenient to use the function of integer argument:

$$F(n) = \frac{(1+\mu)^n - (1-\mu)^n}{4^n}. \tag{2.67}$$

In this case the formula (2.66) will be transformed to the kind:

$$q_{fn} = \frac{1}{\mu}\left(q_{m0}F(n) + 2q_{f0}F(n+1)\right). \tag{2.68}$$

If the number of generations tend to infinity $n \to \infty$, at $\left|\frac{1 \pm \mu}{4}\right| < 1$ and $1 \le \mu < 3$ (since $\mu = 3$ describes the Hardy-Weinberg approach) the limit $\lim\left(\frac{1 \pm \mu}{4}\right)^n \to 0$ if n an integer. In this case, the frequency of the recessive allele a in women tends to zero $q_{fn} \to 0$.

In Figure 2.13 the dependence of the allele a frequency in women from number of generation n is shown at various values of the parameter α plotted under the formula (2.68). The initial frequency of allele a in women was $q_{f0} = 0.7$ and in men $q_{m0} = 0.3$.

From the figure it can be seen that in the absence of incomplete inheritance, i.e., $\alpha = 0$, the allele a frequency in women gradually approaches nearer to the Hardy-Weinberg equilibrium frequency $q_{feq} = \frac{2q_{f0} + q_{m0}}{3} \approx 0.57$ (see also (2.20)). At the occurrence of incomplete inheritance there is an irreversible decrease in the allele a frequency. And if at first a gallop of frequency is observed, then there is a smooth decrease.

Let us find the decrement of the allele a frequency in generations of women:

$$\Delta q_{f(n-1)} = q_{fn} - q_{f(n-1)} = \frac{1}{\mu}\left(\left(q_{m0} - 2q_{f0}\right)F(n) + 2q_{f0}F(n+1) - q_{m0}F(n-1)\right) \tag{2.69}$$

Figure 2.14 shows the calculation of the allele a frequency decrement depending on the number of women in generation n at various values of the parameter α executed under the formula (2.69).

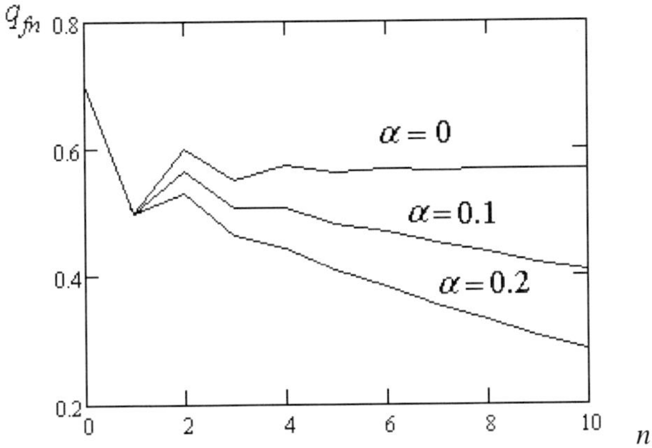

Figure 2.13.

As can be seen from the figure, the approach to the genome equilibrium $\Delta q_{f(n-1)} = 0$ is observed only in the absence of incomplete inheritance, i.e., at $\alpha = 0$.

Otherwise, there is a constant reduction in the allele *a* frequency in generations. And approximately in the seventh generation the constancy of such reduction is established: in the 10th generation at $\alpha = 0.1$ the size $\Delta q_{f(n-1)} = -0.014$, at $\alpha = 0.2$ the size $\Delta q_{f(n-1)} = -0.021$.

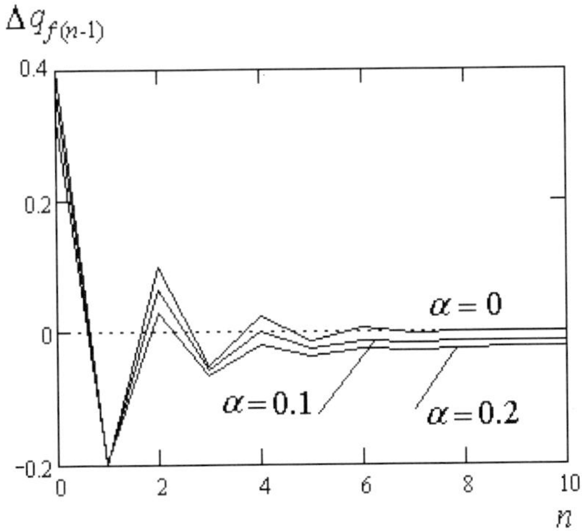

Figure 2.14.

2.7. Inbreeding in a Family Tree

There are family trees and populations in which the principle panmictic mating is essentially broken. Usually, it is in geographically or socially relatively isolated small communities. In these communities the inbreeding - i.e., the consanguineous mating of various orders (a parent - a child, siblings, cousins, uncle - niece, etc.) - can be widespread.

The problem of the influence of inbreeding on the population is the important problem of genetics [3]. To date this problem has not yet received a final solution.

In most cases the inbreeding in a family tree is taken into account in three generations. The most distant relatives who are taken into account regarding consanguineous mating are second cousins.

Inbreeding is taken into account with the help of the *inbreeding factor.*

The inbreeding factor is the probability that two alleles are identical by origin.

Let us consider the method offered by Wright for the calculation of the inbreeding factor in a family tree [3, 11].

For the inbreeding factor it is necessary:

- to find the marrying descendants in all of the ways in a family tree which connect these descendants through the common ancestor.
- to count up the number of steps (alternations of generations) on each part of the way.

Then the inbreeding factor equals:

$$F_t = \frac{1}{2}\sum_{i=1}^{r} 2^{-m_i} \qquad (2.70)$$

where r is the number of ways, and m_1, m_2, ... m_i - the number of steps on each part.

Once again, we shall note that the number of addends in the sum (2.70) is equal to the number of ways through which the common ancestors are connected.

Let us consider concrete cases of family trees (see Figure 2.15). In this figure the square denotes a male individual, and the circle is the female individual. Dotted lines and identical digits specify possible ways in a family tree through which the common ancestor (the man or the woman) connect the descendants through consanguineous marriage. On some sites of ways (Figure 2.15c), there are two digits, 1 and 2. This specifies that such sites are common for the various ways.

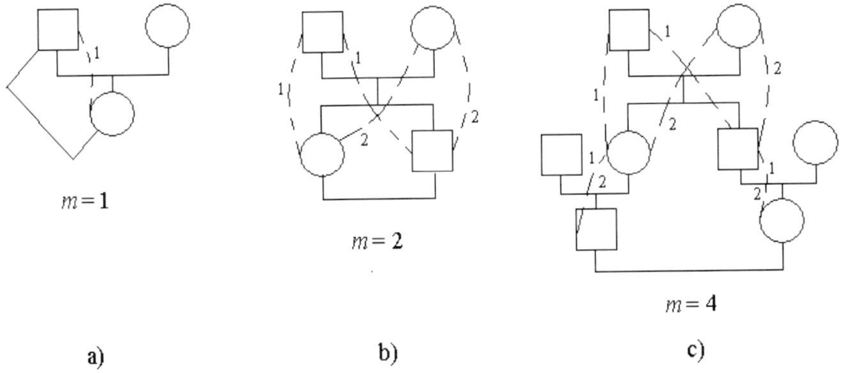

Figure 2.15.

In genetics, marriage is the equivalent of mating between a father and daughter is incest (see Figure 2.15a). The descendant (woman) is connected to the ancestor (man) there 1 way with one step, i.e., $m=1$. Hence, there is an inbreeding factor $F_t=\frac{1}{2}\cdot 2^{-1}=\frac{1}{4}$.

Figure 2.15b shows a marriage between a brother and sister (siblings). The digits specify the number of the way. The numbers of identical digits on the way from the marrying man to the woman is equal to the number of steps.

Let each descendant who has married connect with a partner two ways with two steps each way, i.e., $m=2$.

Hence, there is an inbreeding factor $F_t=\frac{1}{2}\cdot\left(2^{-2}+2^{-2}\right)=\frac{1}{4}$.

The marriage between cousins is shown in Figure 2.15c.

Each descendant who has married connects with a partner in 2 ways, with 4 steps each way, i.e., $m=4$. Hence, there is an inbreeding factor $F_t=\frac{1}{2}\cdot\left(2^{-4}+2^{-4}\right)=\frac{1}{16}$.

The formula (2.70) is easy to understand from the following logic [3].

Obviously, $\frac{1}{2}$ genes of the child are common to each parent, $\frac{1}{4}$ are common to the grandfather and grandmother, $\frac{1}{8}$ to the great-grandfather and great-grandmother, etc. If any ancestor leads the way from the woman with η steps, the share of the common genes with this ancestor equals $2^{-\eta}$. If the man leads the way to any ancestor with μ steps, the share of the common genes equals $2^{-\mu}$. At the marriage of the man and woman with the common ancestor, the share of the common genes with this ancestor will be $2^{-\eta} \cdot 2^{-\mu} = 2^{-m}$ where $m = \eta + \mu$ is the total number of steps through the common ancestor from the woman to the man marrying. The size $\frac{1}{2}2^{-m}$ shows the probability that the gene casually chosen in the woman will be identical in origin to the gene casually chosen in the man. The same size gives the contribution to the inbreeding factor F_t one way through the common ancestor from the woman to the man marrying.

For example, if the common ancestor is $\eta = \mu = 5$ generations away from marrying, the probability of them having the common gene equals $\frac{1}{2}2^{-(5+5)} = \frac{1}{2048}$. This size at the analysis of mating can be neglected.

In Table 6 inbreeding factors for some additional consanguineous marriages are shown.

Table 6.

Type of consanguineous marriage	Inbreeding factor F_t
Second cousins	$\frac{1}{32}$
Cousin uncle - niece	$\frac{1}{32}$
Third cousins	$\frac{1}{64}$

The inbreeding factor F_t is the probability of two alleles in posterity being identical by origin i.e., occurring from the same gene of the ancestor [3]. Or, that the inbreeding factor is a probability of the allele origin in posterity being consanguineous. Hence, the probability of the non-consanguineous origins of identical alleles in homologous chromosomes equals $1 - F_t$.

Let us find the total probability of the recessive homozygote *aa* occurrence (event *aa*) at random mating for autosomal genes [14]:

$$P(aa) = P(H_1)P\left(aa/_{H_1}\right) + P(H_2)P\left(aa/_{H_2}\right). \tag{2.71}$$

This formula follows from the following logic of mating: the origin of identical alleles in homologous autosomes is non-consanguineous (the probability of this hypothesis is $P(H_1) = 1 - F_t$) and there also arises homozygote *aa* (the conditional probability of this event is $P\left(aa/_{H_1}\right) = q^2$); or the origin of identical alleles is consanguineous (the probability of this hypothesis is $P(H_2) = F_t$), and there arises homozygote *aa* (the conditional probability of this event is $P\left(aa/_{H_2}\right) = q$). The probability q of the homozygote *aa* occurrence in the latter case is determined by the fact that if allele a is in one autosome then at the consanguineous origin of the homozygote the same allele necessarily should also be in the homologous autosome.

Hence, the total probability of the recessive homozygote occurrence equals:

$$P(aa) = (1 - F_t)q^2 + F_t q = q^2 + F_t pq, \tag{2.72}$$

where it is taken into account that $p = 1 - q$ is the dominant allele a frequency.

Similarly, there we find the total probability of the dominant homozygote *AA* occurrence at random mating:

$$P(AA) = (1 - F_t)p^2 + F_t p = p^2 + F_t pq \,. \tag{2.73}$$

Thus, according to (2.72) and (2.73) the total frequency of homozygotes (dominant and recessive) in an inbred family tree increases to $2F_t pq$. Hence, the frequency of heterozygotes decreases by the same amount and becomes equal to $2pq - 2pqF_t = 2pq(1 - F_t)$ since the sum of all the zygote frequencies is equal to 1.

In view of the inbreeding factor and also (2.72) and (2.73), the Hardy-Weinberg law in the following generation of the inbred family tree for an autosomal genome (as opposed to (2.2)) needs to be written down with the distributions of zygote genotypes as [7]:

$$(AA)(p_n^2 + F_t p_n q_n) : (Aa)[2p_n q_n (1 - F_t)] : (aa)\left(q_n^2 + F_t p_n q_n\right) \tag{2.74}$$

where p_n is the frequency of the dominant alleles *A* in generation *n*, q_n - the frequency of recessive alleles *a*, so $p_n + q_n = 1$, F_t is the inbreeding factor for a family tree.

The ratio (2.74) also refers to Wright's ratio [6]. Wright's ratio shows that in an inbred family tree after the consanguineous mating of individuals the following effects are observed: there is an increase in the autosomal homozygotes fraction in a family tree on $F_t p_n q_n$ and reduction of the autosomal heterozygotes fraction in $(1 - F_t)$ in comparison with a panmictic population.

But already through one generation the Hardy-Weinberg balance is restored, as it specifies the small importance of the separate consanguineous mating for a family tree:

$$q_{n+1} = \frac{1}{2} 2p_n q_n (1 - F_t) + \left(q_n^2 + F_t p_n q_n\right) = q_n \,. \tag{2.75}$$

Obviously, for a population as a whole the separate act of the consanguineous mating will have still less importance than for a family tree.

The same effect of variation between the frequencies of zygote genotypes is observed for sex-linked genes.

The distribution of the genotypes linked to the *X*-chromosome in a panmictic family tree is described by Hardy-Weinberg law as (2.9).

Let us find the total probability of the occurrence of the recessive homozygote *aa* at random mating for the daughter's genes linked to the *X*-chromosome under the formula (2.71). The logic of events is similar to the case for autosomes except that the conditional probability of the occurrence of a genotype *aa* from the non-consanguineous alleles according to (2.9) equals $P\left(aa/_{H_1} \right) = q_{mn} q_{fn}$. For the consanguineous alleles the conditional probability of a genotype *aa* occurrence equals $P\left(aa/_{H_2} \right) = q_{f(n+1)}$. In daughters, i.e., in generation $n+1$, the probability of a homozygote *aa* occurrence is determined by the fact that if allele *a* is in one *X*-chromosome then at consanguineous origin of the homozygote the same allele necessarily should also be in the homologous *X*-chromosome.

Hence, the total probability of the recessive homozygote occurrence equals:

$$P(aa) = P(H_1) P\left(aa/_{H_1} \right) + P(H_2) P\left(aa/_{H_2} \right) =$$
$$= (1 - F_t) q_{mn} q_{fn} + F_t q_{f(n+1)} = (1 - F_t) q_{mn} q_{fn} + F_t \frac{1}{2} (q_{fn} + q_{mn}) =$$
$$= q_{mn} q_{fn} + \frac{F_t}{2} (q_{fn} + q_{mn} - 2 q_{mn} q_{fn}) = q_{mn} q_{fn} + \frac{F_t}{2} (p_{mn} q_{fn} + p_{fn} q_{mn}) \tag{2.76}$$

where p_{mn} is the frequency of dominant alleles *A* in generation *n* in men, and p_{fn} - in women. For the recessive alleles *a*, accordingly, q_{mn} and q_{fn}. In addition, it is used $p_{mn} = 1 - q_{mn}$ and $p_{fn} = 1 - q_{fn}$.

Similarly, for finding the total probability of the dominant homozygote occurrence with (2.11a):

$$P(AA) = P(H_1)P\left(\frac{AA}{H_1}\right) + P(H_2)P\left(\frac{AA}{H_2}\right) =$$
$$= (1 - F_t)p_{mn}p_{fn} + F_t p_{f(n+1)} = (1 - F_t)p_{mn}p_{fn} + F_t\left[\frac{1}{2}(p_{fn} + p_{mn})\right] =$$
$$p_{mn}p_{fn} + \frac{F_t}{2}(p_{fn} + p_{mn} - 2p_{mn}p_{fn}) = p_{mn}p_{fn} + \frac{F_t}{2}(p_{mn}q_{fn} + p_{fn}q_{mn}) \quad (2.77)$$

and a heterozygote $P(Aa) = 1 - P(aa) - P(AA)$, we shall find the distribution of the genotypes in the *X*-chromosomes of the inbred family tree at mating:

$$(AA)\left[p_{fn}p_{mn} + (p_{mn}q_{fn} + p_{fn}q_{mn})\frac{F_t}{2}\right] : (Aa)[(p_{mn}q_{fn} + p_{fn}q_{mn})(1 - F_t)] : \quad (2.78)$$
$$(aa)\left[q_{mn}q_{fn} + (p_{mn}q_{fn} + p_{fn}q_{mn})\frac{F_t}{2}\right]$$

It is simultaneously demonstrated that as well as in autosomes the inbred additives in dominant and recessive homozygote frequencies in the sex-linked chromosomes are identical.

Another way of the ratio (2.78) finding (2.78) will consist in the following.

Let us first find the total probability of a heterozygote *Aa* occurrence, i.e., $P(Aa)$ at the opportunity of the random consanguineous mating. This probability follows from the following logic of mating: the origin of alleles is non-consanguineous (the probability of this hypothesis is $P(H_1) = 1 - F_t$) heterozygote *Aa* also arises (the conditional probability of this event is $P\left(\frac{Aa}{H_1}\right) = p_{mn}q_{fn} + p_{fn}q_{mn}$, see the formula (2.9)); or the origin of the alleles is consanguineous (the probability of this hypothesis is $P(H_2) = F_t$) and heterozygote *Aa* arises (the conditional probability of this event is equal to zero $P\left(\frac{Aa}{H_2}\right) = 0$). The last statement is defined by the fact that at consanguineous origin similar alleles in homologous *X*-chromosomes, the heterozygote *Aa* cannot arise, consisting different alleles.

Hence, using the formula of total probability we find the total probability of the occurrence of heterozygote Aa:

$$P(Aa) = P(H_1)P\left(\frac{Aa}{H_1}\right) + P(H_2)P\left(\frac{Aa}{H_2}\right) = \tag{2.79}$$
$$= (1-F_t)(p_{mn}q_{fn} + p_{fn}q_{mn}) + F_t \cdot 0 = (1-F_t)(p_{mn}q_{fn} + p_{fn}q_{mn})$$

Reduction of the frequency of heterozygote Aa occurs, as has been proven earlier, due to an identical increase in the frequency of dominant AA and recessive aa homozygotes. For example, the increase in the frequency of recessive aa homozygotes equals:

$$\Delta P(aa) = -\frac{1}{2}\Delta P(Aa) = -\frac{1}{2}\left[-F_t(p_{mn}q_{fn} + p_{fn}q_{mn})\right] = \tag{2.80}$$
$$= \frac{F_t}{2}(p_{mn}q_{fn} + p_{fn}q_{mn})$$

where reduction of the frequency of heterozygotes Aa in a relative of the panmictic family tree according to (2.9) and (2.79) equals $\Delta P(Aa) = -F_t(p_{mn}q_{fn} + p_{fn}q_{mn})$.

Using (2.79) and (2.80) for correcting (2.9), we find the distribution of genotypes (2.78) in X-chromosomes in the daughters of an inbred family tree.

In following the generation, the frequency of the recessive alleles a in women equals:

$$q_{f(n+1)} = \frac{1}{2}(p_{mn}q_{fn} + p_{fn}q_{mn})(1-F_t) + q_{mn}q_{fn} + (p_{mn}q_{fn} + p_{fn}q_{mn})\frac{F_t}{2} =$$
$$= \frac{1}{2}\left[(1-q_{mn})q_{fn} + (1-q_{fn})q_{mn}\right](1-F_t+F_t) + q_{mn}q_{fn} = \frac{1}{2}\left[q_{fn} + q_{mn}\right] \tag{2.81}$$

which fully complies with the same frequency in the absence of inbreeding (2.11), i.e., there is restoration of the Hardy-Weinberg balance; in other words, this is a more exact approach to the balance. Thus, for the genes linked to the X-chromosome the separate act of the consanguineous mating for a family tree has no great importance either.

The ratios (2.75) and (2.81) show as a whole that the allele frequencies in a family tree do not vary on account of inbreeding. There is merely an allele redistribution. A part of the alleles leave the heterozygotes and pass to the homozygotes.

2.7.1. Genetic Load in a Family Tree

In the inbred populations' so-called genetic load, frequently connected to harmful mutations, only recessive alleles passed to offspring are accumulated.

It is connected to increase the frequency of recessive homozygotes *aa* in the inbred populations or family tree, see (2.72) or (2.76), in comparison with panmictic populations (2.2). The increase in the frequency of recessive homozygotes results in an increase in the probability of their mutations.

The concept of the genetic load in the population was used for the first time by the well-known geneticist and Nobel laureate H. J. Muller [12]. However, the theory of the genetic load has actually been developed only for a family tree.

As a result of the increase in the frequency of the recessive homozygtes in the process of consanguineous mating in the population, an increase in their harmful mutations is also observed. Therefore, it is supposed that inbreeding is one of principal causes of the occurrence and accumulation of the genetic load.

The genetic load can be expressed in lethal equivalents. The lethal equivalent is, for example, one lethal mutation resulting in the death of an individual in all cases or two lethal mutations, each of which results in the death of the individuals in 50% cases, etc.

For research of the genetic load in a family tree we shall not take the action of selection into account in order not to complicate the analysis. This is allowable since selection submits to other laws rather than the occurrence of mutations (see paragraphs 2.3, 2.4).

According to the genetic load theory, the formula (2.78) needs to be transformed [3]:

$$\begin{aligned}&(AA)\left[p_{fn}p_{mn}+(p_{mn}q_{fn}+p_{fn}q_{mn})\frac{F_t}{2}\right]:\\&:(Aa)[(p_{mn}q_{fn}+p_{fn}q_{mn})(1-F_t)(1-sh)]:\\&:(aa)\left[q_{mn}q_{fn}+(p_{mn}q_{fn}+p_{fn}q_{mn})(1-s)\frac{F_t}{2}\right]\end{aligned}, \qquad (2.82)$$

where s is the probability of the recessive homozygote destruction as a consequence of the inbreeding, sh - the probability of the heterozygote destruction as a consequence of the inbreeding, h – the so-called dominant degree of the mutation which has occurred owing to inbreeding. If $h=1$, there are heterozygotes and recessive homozygotes perish with the identical frequency, if $h=0$ the heterozygotes are resistant to inbreeding. Dominant homozygotes are assumed to be absolutely resistant to mutations. It is mutational model of process.

Let us notice that the ratio (2.82) concerns a family tree as a separate and usually small population.

Using (2.82), we shall find the recessive allele a frequency in women in the following generation:

$$\begin{aligned}
q_{f(n+1)} &= \frac{1}{2}(p_{mn}q_{fn} + p_{fn}q_{mn})(1-F_t)(1-sh)+\\
&+ q_{mn}q_{fn} + (p_{mn}q_{fn} + p_{fn}q_{mn})(1-s)\frac{F_t}{2} =\\
&= (p_{mn}q_{fn} + p_{fn}q_{mn})\left[\frac{1}{2}(1-F_t)(1-sh)+(1-s)\frac{F_t}{2}\right] + q_{mn}q_{fn} =\\
&= \frac{1}{2}\left[(1-q_{mn})q_{fn} + (1-q_{fn})q_{mn}\right]\left[1 - sh - F_t + F_t sh + F_t - F_t s\right] + q_{mn}q_{fn} =\\
&= \left(\frac{1}{2}(q_{fn}+q_{mn}) - q_{mn}q_{fn}\right)(1 - sh + F_t sh - F_t s) + q_{mn}q_{fn} =\\
&= \frac{1}{2}(q_{fn}+q_{mn}) - \left(\frac{1}{2}(q_{fn}+q_{mn}) - q_{mn}q_{fn}\right)(sh - F_t sh + F_t s) =\\
&= \frac{1}{2}(q_{fn}+q_{mn}) - \frac{1}{2}(q_{fn}+q_{mn})(sh - F_t sh + F_t s) + q_{mn}q_{fn}(sh - F_t sh + F_t s) =\\
&= \frac{1}{2}(q_{fn}+q_{mn})(1 - sh + F_t sh - F_t s) + q_{mn}q_{fn}(sh - F_t sh + F_t s)
\end{aligned} \tag{2.83}$$

In the following generation there is no restoration of the Hardy-Weinberg balance, and hence accumulation of the genetic load in a family tree in a population is observed.

Let us enter the designation $F = sh - F_t sh + F_t s$ - factor of inbreeding for the population. We shall note that if the heterozygotes are resistant to mutations owing to inbreeding $h=0$, and recessive homozygotes perish with probability $s=1$, the inbreeding factors of the populations and family tree equal $F=F_t$. Thus, the formula (2.83) can be written down as:

$$q_{f(n+1)}=\frac{1}{2}\left(q_{fn}+q_{mn}\right)\left(1-F\right)+q_{mn}q_{fn}F\,. \tag{2.84}$$

The lack of the formula (2.84) is the absence of normalization on the general fraction of genes less than 1 which have remained after lethal mutations. However, the standard way of normalization (see paragraph 2.4), results in overly complex formulas which do not allow us to proceed to the analysis of the population.

For updating the formula (2.84), the following logic is used. In the condition of balance of the family tree with inbreeding the formula (2.81) for calculation of the fraction of recessive allele *a* is correct. The formula (2.81) is absolutely correct since it is normalized on the general portion of genes equal to 1.

Taking into account (2.81), and also $F<<1$ [3] and the small size of the second addend in the right part (2.84), it is possible to assume that for normalization (2.84) it is necessary to divide the right part into the value $1-F$.

Therefore, the corrected formula (2.84) looks like:

$$q_{f(n+1)}=\frac{1}{2}\left(q_{fn}+q_{mn}\right)+q_{mn}q_{fn}\frac{F}{1-F}\,. \tag{2.85}$$

The formula (2.85) allows us to find the recessive allele *a* fraction in women in the following generation of the family tree as the sum of equilibrium addend and gradually increasing owing to inbreeding in the genetic load.

We shall carry out the further analysis of a genetic load accumulation in populations in paragraph 3.6.1.

Chapter 3

3. MATHEMATICAL GENETICS OF POPULATION

The Hardy-Weinberg law of genotype distribution in generations for the autosomal genome (2.2) or for genome linked to the X-chromosome (2.9) which was found by English mathematician Hardy and German physicist Weinberg in 1908 plays a key role in the mathematical analysis of genetic processes. Founders of the law assumed, that in the given kind it is correct for populations.

However, it is not absolutely so. The point is that the population will consist of family trees set which periodically contact among themselves. In Figure 3.1 the principle of a population formation from separate family trees is shown. The square means a male individual; the circle represents a female.

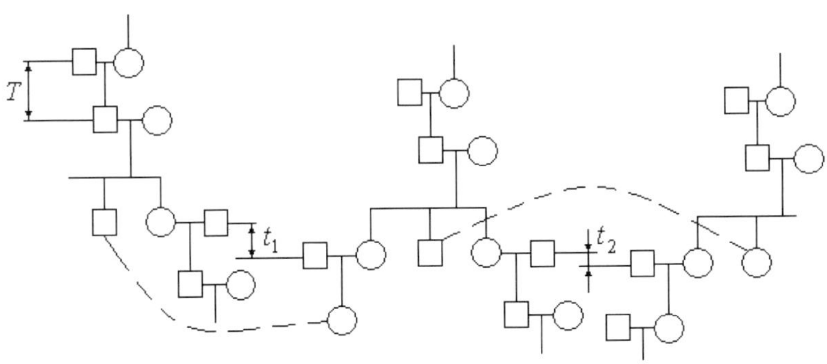

Figure 3.1.

As an example, three family trees are shown. If, for a family tree, it is possible to consider the time of one generation alternation as approximately $T \approx 25 - 30$ years for a population, this time can be any t_1, t_2, etc. For a population the continuous alternation of generations is characteristic and hence the form of the Hardy-Weinberg law record should not have a discrete character. In it is essence of the basic distinction of the Hardy-Weinberg law record for a population and a family tree. In earlier used form of the Hardy-Weinberg law (2.2) or (2.9) this law is written down for a family tree.

3.1. HARDY: WEINBERG LAW FOR A POPULATION

As was already specified in the form considered above, the Hardy-Weinberg law concerns a separate family tree. Implicitly, this law includes time, since the alternation of generations occurs within a certain time *T*. Thus, the Hardy-Weinberg law, for example, such as (2.4) for genes of the autosome or (2.11) for genes of the *X*-chromosome allowing us to find the frequency of a recessive allele in the following generation for the expressed discrete character on time.

The population will consist of family trees crossed among themselves and lives in continuous time. Alternation of generations of a family trees set results to that generations vary actually according to a continuous time scale.

The further analysis in this paragraph concerns to a panmictic populations.

Writing of the Hardy-Weinberg law (2.4) reflecting preservation of the autosomal genome allele frequencies for a population is trivial therefore the big interest does not represent. It expresses constancy of the allele frequency in time:

$$q_n = const\,. \tag{3.1}$$

In (3.1) the continuous time scale *n* is used. Under size $n = \frac{t}{T}$ in this case it means the current time of a population life *t* normalized on average in family trees of a population the time *T* of one generation life, i.e., actually continuous, dimensionless time. For the convenience of transformations as opposed to the usual way of function writing, we shall further use a record where the argument of function is designated as an index $q(n) = q_n$. The distinction between dimensionless time and the number of the generation will be clear from the context and will not cause difficulty. Before beginning to read the following text, it is recommended to repeat paragraph 2.1.4.

It is more informative rather than (3.1) the Hardy-Weinberg law (2.11), reflecting the inheritance of the genes linked to the *X*-chromosome.

Let us consider the differential equation:

$$\frac{d^2 q_{f(n-1)}}{dn^2} + \eta \frac{dq_{f(n-1)}}{dn} = 0, \tag{3.2}$$

where η is constant. In this formula $q_{f(n-1)}$ there is a frequency of recessive alleles in the female in generation $n-1$. Differentiation in (3.2) goes on dimensionless time n.

Let us transform the equation (3.2) to finite-differential form:

$$\frac{q_{fn} - 2q_{f(n-1)} + q_{f(n-2)}}{\Delta n^2} + \eta \frac{q_{f(n-1)} - q_{f(n-2)}}{\Delta n} = 0, \tag{3.3}$$

where size Δn is the time step determining the accuracy of finite-differential representation of a derivative.

Uniting similar members and multiplying the equation (3.3) on -2 we shall find:

$$-2q_{fn} + 2(2 - \eta\Delta n)q_{f(n-1)} - 2(1 - \eta\Delta n)q_{f(n-2)} = 0. \tag{3.4}$$

Let us try to identify the equations (3.4) and (2.14) for alternation of the recessive allele frequencies in generations of a family tree. For this purpose it is necessary to accept:

$$2(2 - \eta\Delta n) = 1 \text{ and } -2(1 - \eta\Delta n) = 1. \tag{3.5}$$

The wonderful feature of the algebraic equations (3.5) is that they have an identical solution:

$$\eta \Delta n = \frac{3}{2} \cdot \tag{3.6}$$

This means that the finite-differential equation (2.14) and the differential equation (3.2) can have the same characteristic equation. Taking into account (3.6), the equation (3.2) can be copied as:

$$\frac{d^2 q_{f(n-1)}}{dn^2} + \frac{3}{2\Delta n} \frac{dq_{f(n-1)}}{dn} = 0 \cdot \tag{3.7}$$

The equation (3.7) can be integrated once:

$$\frac{dq_{f(n-1)}}{dn} + \frac{3}{2\Delta n} q_{f(n-1)} = C_1 \cdot \tag{3.8}$$

where C_1 is the constant of integration.

Further integrating the equation (3.8) by the method of separation of variables:

$$\int\limits_{q_{f0}}^{q_{fn}} \frac{dq_{f(n-1)}}{C_1 - \frac{3}{2\Delta n} q_{f(n-1)}} = \int\limits_0^n dn \,, \tag{3.9}$$

where q_{f0} is the initial examined frequency of the recessive allele in the woman, we shall find $q_{fn} = q_f(n)$ - time dependence of the allele frequency:

$$q_{fn} = \frac{2\Delta n}{3} C_1 - \left(\frac{2\Delta n}{3} C_1 - q_{f0} \right) e^{-\frac{3}{2\Delta n} n} . \tag{3.10}$$

Identifying the solution (3.10) with the solution (2.20), we shall find:

$$\frac{2q_{f0}+q_{m0}}{3}=\frac{2\Delta n}{3}C_1 \text{ and } \frac{q_{f0}-q_{m0}}{3}=-\left(\frac{2\Delta n}{3}C_1-q_{f0}\right) \qquad (3.11)$$

where q_{m0} is the initial examined frequency of the recessive allele in the man.

As expected, the formulas (3.11) do not contradict each other. Hence, the solution (3.10) can be written down as:

$$q_{fn}=\frac{2q_{f0}+q_{m0}}{3}+\left(\frac{q_{f0}-q_{m0}}{3}\right)e^{-\frac{3}{2\Delta n}n}. \qquad (3.12)$$

The formula (3.12) is correct for the frequency of female allele a only in even generations. This is a consequence of the transition to the continuous scale of time n.

Comparing (2.20) and (3.12), for even generations we have $2^{-n}=e^{-\frac{3}{2\Delta n}n}$ or $\Delta n=\frac{3}{2\ln 2}$. Hence, (3.12) will be transformed to the kind:

$$q_{fn}=\frac{2q_{f0}+q_{m0}}{3}+\left(\frac{q_{f0}-q_{m0}}{3}\right)e^{-n\ln 2}, \qquad (3.13)$$

which is identical to the formula (2.20) for even generations n. But in (3.13), as opposed to (2.20), the size n is a dimensionless time. Numbers of generation in (3.13) are not present.

Taking into account (3.6) and $\Delta n=\frac{3}{2\ln 2}$ we shall find $\eta=\ln 2$. Thus, the differential equation (3.2) will be written down as:

$$\frac{d^2 q_{f(n-1)}}{dn^2}+\ln 2\frac{dq_{f(n-1)}}{dn}=0. \qquad (3.14)$$

As we used the continuous dimensionless time n the concept of generation numbers which was still kept in (3.2) actually does not play any role and it is possible to write the formula (3.14) down as:

$$\frac{d^2 q_f}{dn^2} + \ln 2 \frac{dq_f}{dn} = 0. \tag{3.15}$$

The differential equation (3.15) is the Hardy-Weinberg law in case of the continuous alternation of generations for a panmictic population, i.e., for the continuous time scale in the absence of mutagen influence on the population. The equation (3.15) defines the time dependence of recessive allele a frequency $q_f(n)$ linked to the X-chromosome for women in the population.

The differential equation (3.15), as it is determined by the Hardy-Weinberg law, reflects the indifferent equilibrium of genome (see paragraph 2.1.2). Really, the equation (3.15) can be satisfied by the solution $q_f = const$, i.e., any constant value of allele frequency is stable.

Let us note that in connection with the existence of the ratio $q_f = 1 - p_f$ such an equation can be written down and for the frequency of female dominant alleles A it can be linked to the X-chromosome in a population:

$$\frac{d^2 p_f}{dn^2} + \ln 2 \frac{dp_f}{dn} = 0. \tag{3.15a}$$

3.2. Mutagenesis in a Population

Mutations are an extremely important topic of mathematical genetics (see also paragraph 2.2). The mutations explain genetically dependent hereditary diseases. They allow us to explain some of the processes of the evolution of organisms. Mutations underlie the breeding of animals and plants.

Mutations usually are spontaneous; they arise spontaneously and are induced, arising under the action of an external factor, for example, radiation [6].

During mutation the mutagen factor has the important role. It is possible to classify mutagen factors into two groups: determined and stochastic.

Let us consider the action of a stochastic mutagen factor on a population.

3.2.1. The Stochastic Mutagen Factor: Newgrowths

Let us analyze the influence of the stochastic mutagen factor on the example of the occurrence of malignant newgrowths.

The problem of the population death rate from oncological diseases is very many-sided and includes questions of early diagnostics of diseases, effective and accessible treatment, rehabilitation actions, etc. More than 6 million humans around the world die annually from oncological diseases [15].

Among other kinds of diseases, the occurrence of malignant newgrowths has some distinctive features. First of all, there is a great variability in the site of the newgrowths. It can arise in practically any part of the organism. In addition, for oncological diseases there are typically a variety of mutagen (cancerigenic) factors: poor-quality food, polluted environment, mode of life and professional work, smoking and many other factors.

All of these cancerogenic factors finally affect the mitogenetic function of a cell, causing its malignant transformation.

Generalizing, we shall consider that the set of reasons resulting in the occurrence of the malignant newgrowths results from the influence on the organism of some stochastic mutagen factor.

Despite the stochastic character of the influence, it is difficult to assume the situation in which the given stochastic mutagen factor would be completely absent. It concerns even completely isolated primitive societies. In particular, such a kind of factor is always present in a modern, civilized society.

As a model for investigation we shall consider a homogeneous and stable human society of a very much advanced country with a high level of medical development which is accessible to all of the population. In such countries the death rate of the population should basically be caused by oncological diseases which start to play the role of the natural factor of inevitable alternation of generations. We shall name such countries' demographic 'stationary'.

It is possible to assume also that in similar societies the genetic-mathematical laws determining the death rate of the population from oncological diseases should operate, i.e., as a result of the action of the stochastic mutagen factor.

To understand how it is possible to take into account the action of the stochastic mutagen factor, we will address another well-investigated physical phenomenon –the Brownian motion [16].

Brownian motion is the continuous spontaneous chaotic movement of microparticles (Brownian particles in diameter of less than one micrometer) weighed in a liquid or gas. This movement is caused by chaotic impacts of molecules of an environment owing to their thermal movement. This was described for the first time by English botanist R. Brown in 1828.

Brownian motion of a particle in a liquid at first glance should not exist. Really, according to the Brownian particle, for example, flower pollen impacts the molecules of the liquid whose operation is counterbalanced from different directions. Therefore, the most probable condition of the particle is motionless. The particle should shiver only but should not have some constant displacement from a point of supervision. The outstanding scientists A. Einstein and M. Smoluchowski have shown that physically the Brownian motion is a consequence of the statistical properties of the second law of thermodynamics. If the researcher has a relatively small number of molecules, the essential deviation from the most probable state of the system should be observed, in this case the motionless state of the Brownian particles.

Let us note the main similarity between two phenomena: the Brownian motion and the existence of the population in conditions of the stochastic mutagen factor action.

At the Brownian motion on the determined system - particle in the liquid – stochastic force acts from the molecules of a liquid.

In the researched case on the determined system - reproductive genome – some stochastic mutagen factor acts.

At the Brownian motion the equation of the movement of the particle looks like [16]:

$$m\frac{d^2S}{dt^2}+r\frac{dS}{dt}=F\,, \tag{3.16}$$

where m is the mass of the particle, S - the displacement of the particle from the initial position, r – the factor of medium resistance to the movement of the particle, t - time, F – the stochastic force acting on the particle from the molecules of liquid. We shall note the absence in the equation (3.16); the elastic force, which is determined, returned the particle to its initial position, causing it to oscillate around the balance point.

The equation (3.16) is the equation which – in the absence of stochastic function F - satisfies the solution $S = const$. I.e., at $F = 0$ the particle can steadily be in any position - indifferent balance.

The equation (3.15) is similar to the equation (3.16) for $F = 0$.

In this connection we shall note that function $q_f = const$ satisfies the equation (3.15), i.e., the frequency of allele a in the X-chromosome of the woman is constant for any value that reflects the steady-indifferent character of the Hardy-Weinberg balance for the sex-linked genome. If there is some stochastic mutagen factor $D(n)$ in the randomly time-dependent lives of the population (in conditions of a continuous time scale of alternation of generations), the equation (3.15) by analogy with (3.16) needs to be copied as:

$$\frac{d^2 q_f}{dn^2} + \ln 2 \frac{dq_f}{dn} = D(n). \tag{3.17}$$

Using the result found for the first time by A. Einstein [16] for the Brownian motion $\left\langle (\Delta S)^2 \right\rangle \sim t$, we shall note that the average square of the deviation from the norm of the frequency of the allele a (3.13) at the action on the population of the stochastic mutagen factor is proportionally the time of the population life $\left\langle \left(\Delta q_f\right)^2 \right\rangle \sim n$. Angular brackets mean the average for individuals of the population.

Thus, during the life of the population at the action of the stochastic mutagen factor, the mean square deviation from the norm of the allele frequency is proportionally to the root square from the time of the population life $\sqrt{\left\langle \left(\Delta q_f\right)^2 \right\rangle} \sim \sqrt{n}$. At the certain level of the mean square deviation from the norm of the allele frequency can lead to a lethal outcome. For each separate individual the lethal deviation is individual.

The received result shows that during the population life and the alternation of generations at the action of the stochastic mutagen factor the death rate inevitably grows (similarly to displacement of the Brownian particles from the point of initial supervision). This conclusion has a completely general biology-mathematical character and is also a consequence of the second law of thermodynamics, i.e., a consequence of the inevitable growth of entropy in the population.

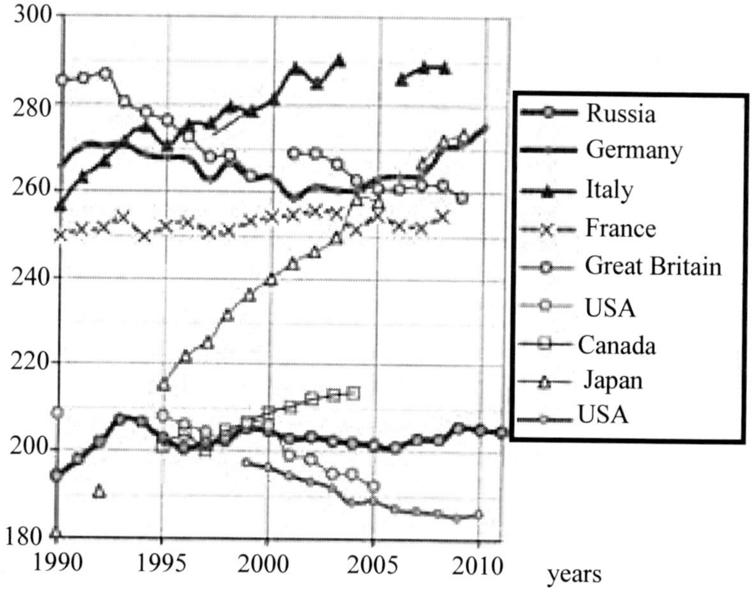

Figure 3.2.

In Fig 3.2 the dynamics of death rates (mortality rate coefficient) of the population in the various countries from newgrowths is shown [17].

A mortality rate coefficient it is the ratio of the number of people who died in the country for each year in proportion to the average number of the people in this country and then multiplied by 1000.

The time interval of 20 years during which the death rate was investigated is a short term but it is possible to draw some conclusions.

In two countries, Japans and Canada, the parabolic law: death rate $\sim \sqrt{n}$ is obviously observed. The distinctive features of these countries include the very high level of medicine and the great uniformity of the population, which almost without exception, uses these achievements of medicine. Some other social factors determining on the whole a positive psychological climate in these countries are also influential. In other words, the situation with the detection at the earliest stage and treatment of the newgrowths in these countries has approached the stationary limit on the given level of development of the country. The lack of changes in this direction, therefore the law-death rate proportional $\sim \sqrt{n}$, is carried out. The further decrease in the death rate will take place at the occurrence and universal application of essentially new methods of diagnostics and the treatment of cancer. The similar development of medical science demanding big financial expenses and intellectual efforts is counteracting the growth entropy in the population. In the given countries, the death rate from newgrowths is the basic natural factor of alternation of the generations.

For other countries the first characteristic is the great heterogeneity of the population; second is the high immigration of a foreign population which gradually joins modern medicine, which leads to a reduction in the general death rate from newgrowths (USA, Great Britain) or to an invariance (Germany, Russia, France). As a whole it is possible to speak about a generally unsteady demographic in these countries. Pay attention to the example of Italy, for which the law - death rate $\sim \sqrt{n}$ is carried out with periodic fluctuations. Apparently, it is connected by the fact that Italy is basically the transit country for the foreign population. However there is periodical delay of this people in the country results in the fluctuations of a mortality rate coefficient on a background of the death rate law $\sim \sqrt{n}$.

In conclusion of the analysis we shall note that the occurrence of newgrowths basically occurs in people of old age. The average life duration in Russia, for example, is much less than in other countries of Europe and in Japan. The proportion of older people is also smaller. Therefore, the death rate from newgrowths in Russia is also less than in other countries of Europe and Japan (Figure 3.2). It is known that the death rate in Russia is basically the consequence of diseases of the cardiovascular system. Thus, the alternation of generations occurs basically not due to natural, but to social factors.

3.2.2. Spontaneous Mutagenesis in a Population: Hemophilia

The occurrence of hemophilia is usually related to *spontaneous* mutations. However, as was already explained in paragraph 2.3, it is possible to examine the problem of hemophilia assuming the action of some equivalent determined constant mutagen factor. If this mutagen factor constantly did not operate, illness due to selection would quickly disappear.

Let us consider the action of the constant mutagen factor on the population of two-alleles genome linked to the X-chromosome.

We shall make the analysis on the basis of the Hardy-Weinberg law for the population (3.15) which is written as:

$$\frac{d^2 q_f}{dn^2} + \ln 2 \frac{dq_f}{dn} = \alpha, \tag{3.18}$$

where q_f is the frequency of the recessive allele in the *X*-chromosome in the woman, and value α - characterizes some equivalent constant mutagen factor causing hemophilia in a population.

The equation (3.18) can be integrated once:

$$\frac{dq_f}{dn} + \ln 2 q_f = \alpha n + C_3, \tag{3.19}$$

where C_3 is the constant of integration.

The equation (3.19) is integrated in quadratures. The general solution looks like:

$$q_f = \frac{\alpha n}{\ln 2} - \frac{\alpha}{(\ln 2)^2} + \frac{C_3}{\ln 2} + C_4 e^{-n \ln 2}, \tag{3.20}$$

where C_4 is the second constant of integration.

In connection with the fact that the basic results of the genetic research into hemophilia have been earlier received for a family tree (see paragraph 2.3), for the use of the given principle of discrete alternation generations at the analysis stage for the use of the previous researchers' results it is convenient to return to the discrete scale of generations.

The frequency of the variation of alleles *a* in the woman for one generation equals:

$$\Delta q_{f(n-1)} = q_{fn} - q_{f(n-1)} = \frac{\alpha(n+1)}{\ln 2} - \frac{\alpha}{(\ln 2)^2} + \frac{C_3}{\ln 2} + C_4 e^{-(n+1)\ln 2} - \left(\frac{\alpha n}{\ln 2} - \frac{\alpha}{(\ln 2)^2} + \frac{C_3}{\ln 2} + C_4 e^{-n \ln 2} \right) = \frac{\alpha}{\ln 2} - \frac{1}{2} C_4 e^{-n \ln 2} \tag{3.21}$$

At the increase in the number of generations $n \to \infty$ the change of frequency of the alleles *a* is $\Delta q_{f\infty} = \frac{\alpha}{\ln 2}$.

The frequency of mutations *de novo* is determined as the ratio of the number of the anomaly display events to the double number of the surveyed individuals [3] $\mu = \frac{m}{2n}$, where m is the number of the occurrence of the given anomaly events (phenotypic displays of mutation) equal to the number of individuals having the mutant allele, n – the number of the surveyed individuals (men and women).

Taking into account the determination of the allele frequency, see paragraph 2.1.2, we find an interrelationship between the change of the frequency of the recessive allele *a* in women and the frequency of the mutations $\mu = \frac{\Delta q_f}{2}$. For calculation it is convenient to use the change of allele *a* frequency at the alternation of the big number of generations $\Delta q_{f\infty} = \frac{\alpha}{\ln 2}$. In this case we find an interrelationship between the equivalent constant mutagen factor and the frequency of mutations $\alpha = 2\mu \ln 2$.

The frequency of the mutations for hemophilia in different countries (into the population) varies from $\mu = -2.2 \cdot 10^{-5}$ (Switzerland) up to $\mu = -3.2 \cdot 10^{-5}$ (Denmark) [3]. The minus sign reflects the reduction in the frequency of healthy genes at mutations $\Delta q_{f\infty} < 0$.

Let us state that at $n \to \infty$ the size is $\mu \approx -3 \cdot 10^{-5}$. In this case the equivalent constant mutagen factor is $\alpha = 2\mu \ln 2 \approx -4.16 \cdot 10^{-5}$.

The first uncertain size in the dynamics of change of the allele frequencies (3.21) is the constant C_4. According to (3.21) we shall find the law of genic frequency decrease:

$$q_{fn} = q_{f(n-1)} + \frac{\alpha}{\ln 2} - \frac{1}{2} C_4 e^{-n \ln 2}. \tag{3.22}$$

We use the initial condition: for $n=1$ according to (2.11) $q_{f1}=\frac{q_{m0}+q_{f0}}{2}$.

Hence, using (3.22) we shall find:

$$q_{f1}=q_{f0}+\frac{\alpha}{\ln 2}-\frac{1}{4}C_4=\frac{q_{m0}+q_{f0}}{2}. \tag{3.23}$$

From (3.23) we find constant C_4:

$$C_4=2\left(q_{f0}-q_{m0}\right)+\frac{4\alpha}{\ln 2}. \tag{3.24}$$

Substituting (3.24) in (3.21) we shall find:

$$\Delta q_{f(n-1)}=\frac{\alpha}{\ln 2}-\left(\left(q_{f0}-q_{m0}\right)+\frac{2\alpha}{\ln 2}\right)e^{-n\ln 2}. \tag{3.25}$$

Believing for definiteness of calculation $q_{f0}=q_{m0}$, we shall receive:

$$\Delta q_{f(n-1)}=\frac{\alpha}{\ln 2}\left(1-2e^{-n\ln 2}\right)=\frac{\alpha}{\ln 2}\left(1-2^{-(n-1)}\right). \tag{3.26}$$

For $n=0$ we find the initial change of alleles O frequency which equals $\Delta q_{f(-1)}=-\frac{\alpha}{\ln 2}$. Taking into account $\alpha\approx -4.16\cdot 10^{-5}$ we find $\Delta q_{f(-1)}\approx 6\cdot 10^{-5}$.

We shall note that the value $\Delta q_{f(-1)}$ has a rated character. The change of genic frequency at mutagenesis is real and begins from the time coordinate $n=1$ at which, according to (3.26), is $\Delta q_{f(n-1)}=\Delta q_{f(0)}=0$. (See Figure 3.3 and Figure 3.4.)

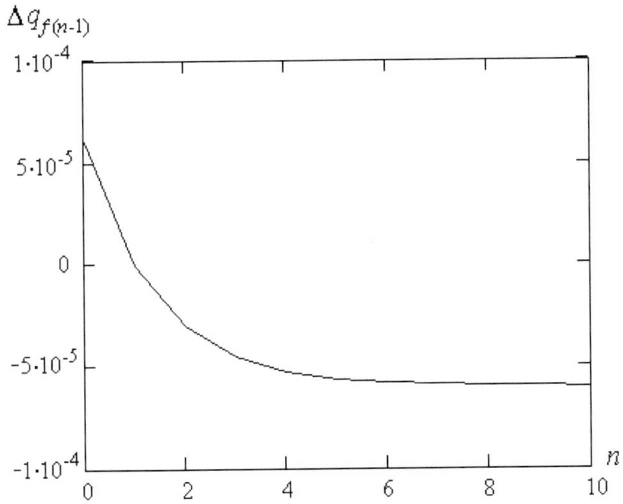

Figure 3.3.

In Figure 3.3 the change in the dependence of genic alleles O frequency $\Delta q_{f(n-1)}$ in women on the dimensionless time of the population life plotted under the formula (3.26) for hemophilia is shown.

The analysis of Figure 3.3 allows us to assume that under the action of the equivalent constant mutagen factor α there is an average of individuals of the population with the mutation (practically constant after $n > 8$) resulting in an average among the population with continuous reduction of healthy alleles O frequency in women and also an increase over time in the frequency of destructive alleles O i.e., to the hemophilia.

The law (3.20) of decrease in the frequency of blood alleles O in mutations with the account (3.24) and $q_{f0} = q_{m0}$ can be written down as:

$$q_f = \frac{\alpha n}{\ln 2} - \frac{\alpha}{(\ln 2)^2} + \frac{C_3}{\ln 2} + \frac{4\alpha}{\ln 2} e^{-n\ln 2}. \tag{3.27}$$

Using the initial condition: for $n = 0$, $q_f = q_{f0}$ we shall find:

$$C_3 = q_{f0} \ln 2 + \frac{\alpha}{\ln 2} - 4\alpha. \tag{3.28}$$

Taking into account (3.28), we shall write the formula (3.27) as:

$$q_f = q_{f0} + \frac{\alpha n}{\ln 2} - \frac{4\alpha}{\ln 2}\left(1 - e^{-n\ln 2}\right). \tag{3.29}$$

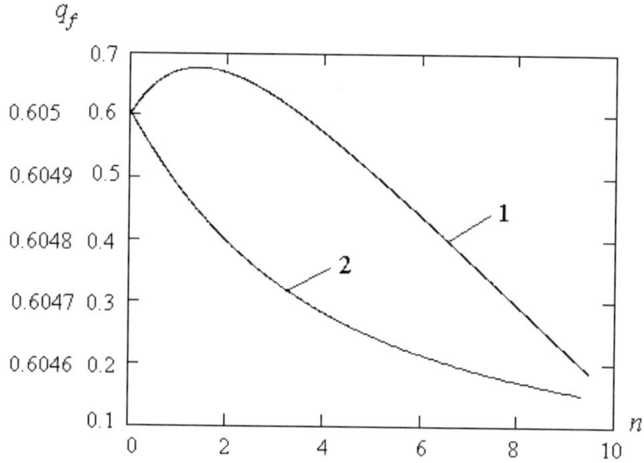

Figure 3.4.

In Figure 3.4 (curve 1, the left scale of the ordinates axis) the graph of dependence of the decrease in allele blood O frequency q_f on dimensionless time n of the population life with hemophilia plotted under the formula (3.29) for initial allele frequency $q_{f0} = 0.605$ [3]; see also formula (2.8).

3.3. Frequency of Mutations and the Equivalent Mutagen Factor for a Family Tree with Hemophilia

Let us find the interrelationship between the equivalent constant mutagen factor used for the analysis of a family tree b, see paragraph 2.3, and the frequency of mutations μ. Comparing formulas (3.20) and (2.29), we find $\frac{\alpha}{\ln 2} = \frac{2}{3}b$; or taking into account $\alpha = 2\mu \ln 2$, we have $b = 3\mu$.

In paragraph 2.3 the graph in Figure 2.6 is plotted for the frequency of mutations $\mu = -6.5 \cdot 10^{-5}$.

Let us note that the formula $b = 3\mu$ for a family tree has no great practical value. Really, the equivalent constant mutagen factor for the calculation of hemophilia in a family tree is not usually connected with mutagenesis *de novo* and the size of this factor has absolutely other order of magnitude than for a population. In paragraph 2.3, in the analysis of hemophilia, for example, in the European royal houses (Figure 2.8), for a reflection of the real picture of the disease the equivalent constant mutagen factor was accepted as being equal to $b = 0.178$.

Let us analyze this size in more detail. Let us imagine that in a family tree there are $2n$ individuals in which there are zygotes $3n$ *X*-chromosomes (two *X*-chromosomes in a female zygote and one in a male zygote).

Naturally, the number of male hemophiliacs is equal to the number of the *X*-chromosomes damaged by the mutation in their zygotes δn, where δ there is the average proportion of male hemophiliacs in a family tree. Eliminated from a family tree $(1-f)\delta n$ of the hemophilia genes [3], where f – fertility of male hemophiliacs, i.e., their probability of leaving posterity. In the case of the balance of mutations and selection, the mutations in those men who do not leave posterity should result in the replacement of a certain number of normal genes by mutant genes.

Product $(1-f)\delta$ is the proportion of male hemophiliacs not leaving posterity. The frequency of mutant alleles in a family tree falling one *X*-chromosome in a family tree equals $\mu_t = \frac{(1-f)\delta n}{3n} = \frac{(1-f)\delta}{3}$, where μ_t is the frequency of mutations for a family tree (not equal to μ). If we were to use the ratio we should find $b = (1-f)\delta$, i.e., the equivalent constant mutagen factor in a family tree is equal to the proportion of male hemophiliacs not leaving posterity.

Hence, for the dynasty shown in Figure 2.8 the equivalent constant mutagen factor for generation $n = 1$ is equal to zero, and for generation $n = 2$ is equal to $b = \frac{3}{7} = 0.43$.

Believing that at the alternation of a large number of generations in a family tree the equivalent mutagen factor should aspire to a constant

$b \rightarrow const$, for calculation of the dynasty shown in Figure 2.8 it is possible to take the average value of the mutagen factor for these two generations $b = \frac{3}{14} = 0.214$.

For this size of equivalent mutagen factor the graph of dependence for hemophilia of the recessive allele O frequency q_{fn} in the *X*-chromosome in women in the number of generation is shown in Figure 2.9.

3.4. Natural Select in a Population

For the analysis of the natural selection action on a population, see also paragraph 2.4; for the purpose of using the previous researches we shall return to a separate family tree. In the family tree where hemophilia is observed, the selection resulting in a decrease in genic frequencies - in particular, of the allele O blood - operates.

The action of selection is intensive enough. For example, the life duration of men who suffer from hemophilia reaches 1/3 of the life duration of healthy people [3], so male fertility (i.e., the chances of having posterity) in comparison to healthy men is reduced. Therefore, not all men of the given family tree participate in the reception of posterity and damaged alleles O are eliminated from the family tree.

Let us consider selection in the case of homozygotes *aa* (or OO).

Genotypes before selection, for example, in generation $n-2$, are distributed according to (2.9).

We accept the ratio of genotype fitnesses, see also paragraph 2.4:

$$1:1:(1-s), \tag{3.30}$$

where s is a reduction of the homozygotes fraction of the recessive allele a as a result of selection (paramcter of selection).

We shall write down genotypes after selection for the following generation $n-1$:

$$(AA)\, p_{f(n-1)} p_{m(n-1)} : (Aa)\left(p_{m(n-1)} q_{f(n-1)} + p_{f(n-1)} q_{m(n-1)}\right) :$$
$$: (aa)\, q_{m(n-1)} q_{f(n-1)} (1 - s)$$

(3.31)

Taking into account that

$$p_{f(n-1)} p_{m(n-1)} + \left(p_{m(n-1)} q_{f(n-1)} + p_{f(n-1)} q_{m(n-1)}\right) + q_{m(n-1)} q_{f(n-1)} =$$
$$= \left(p_{f(n-1)} + q_{f(n-1)}\right)\left(p_{m(n-1)} + q_{m(n-1)}\right) = 1$$

we find the sum of the genotypes frequencies in formula (3.31): $1 - q_{m(n-1)} q_{f(n-1)} s$.

Furthermore, using a standard rule of an allele frequency finding in the following generation (half of heterozygotes frequency plus that of homozygotes frequency) and the formula (3.31), we calculate the frequency of recessive alleles a in women in generation n for sex-linked genes which is similar [3] where such calculation is made for autosomal genome:

$$q_{fn} = \frac{\frac{1}{2}\left(p_{m(n-1)} q_{f(n-1)} + p_{f(n-1)} q_{m(n-1)}\right) + q_{m(n-1)} q_{f(n-1)} (1 - s)}{1 - q_{m(n-1)} q_{f(n-1)} s}$$

(3.32)

Let us transform the formula (3.32) using $p_{m(n-1)} = 1 - q_{m(n-1)}$ and $p_{f(n-1)} = 1 - q_{f(n-1)}$:

$$q_{fn} = \frac{\frac{1}{2}\left(q_{f(n-1)} + q_{m(n-1)}\right) - q_{m(n-1)} q_{f(n-1)} s}{1 - q_{m(n-1)} q_{f(n-1)} s}. \tag{3.33}$$

We believe that the frequency of allele a in the men is equal to the frequency of allele a in the women of the previous generation $q_{m(n-1)} = q_{f(n-2)}$. Hence:

$$q_{fn} = \frac{\frac{1}{2}\left(q_{f(n-1)} + q_{f(n-2)}\right) - q_{f(n-2)} q_{f(n-1)} s}{1 - q_{f(n-2)} q_{f(n-1)} s}. \quad (3.34)$$

Mutagenesis and selection operate in one direction. If they have arisen on account of mutagenesis the damaged alleles will have the same velocity due to selection eliminating some from a family tree to maintain the balance. Some quantity of the damaged genes are kept in the family tree but this quantity will not increase.

It is possible to calculate the variation of the frequency of allele *a* in the family tree using (3.34) under the formula:

$$\Delta q_{f(n-1)} = q_{fn} - q_{f(n-1)} = \frac{\frac{1}{2}\left(q_{f(n-1)} + q_{f(n-2)}\right) - q_{f(n-2)} q_{f(n-1)} s}{1 - q_{f(n-2)} q_{f(n-1)} s} - q_{f(n-1)} =$$

$$= \frac{\frac{1}{2}\left(-q_{f(n-1)} + q_{f(n-2)}\right) - q_{f(n-2)} q_{f(n-1)} s + q_{f(n-2)} q_{f(n-1)}^2 s}{1 - q_{f(n-2)} q_{f(n-1)} s}$$

(3.35)

At the transition to the continuous time scale we believe the difference in the frequencies of two consecutive generations is infinitesimal, i.e., $q_{f(n-2)} \approx q_{f(n-1)}$. Hence, the formula (3.35) will be transformed to the similar kind of autosomal genome [3]:

$$\Delta q_{f(n-1)} = -\frac{q_{f(n-1)}^2 s\left(1 - q_{f(n-1)}\right)}{1 - q_{f(n-1)}^2 s}. \quad (3.36)$$

For the transition from a family tree to a population we shall take the method offered in [3] for the calculation of allele *a* frequency in a large number of generations. We shall copy (3.36) as:

$$\frac{dq_f}{dn} = -\frac{q_f^2 s\left(1 - q_f\right)}{1 - q_f^2 s}. \quad (3.37)$$

We do not write the number of the generation in the index of allele frequency for the population, n it is dimensionless time. We shall integrate (3.37):

$$\int_{q_{f0}}^{q_f} \frac{dq_f}{q_f^2 s(1-q_f)} - \int_{q_{f0}}^{q_f} \frac{dq_f}{(1-q_f)} = -\int_0^n dn \cdot \qquad (3.38)$$

As a result, we shall find:

$$\frac{q_{f0}-q_f}{q_{f0}q_f} + \ln \frac{q_{f0}}{q_f}\left(\frac{1-q_f}{1-q_{f0}}\right)^{1-s} = sn \cdot \qquad (3.39)$$

For the blood system ABO, as before the initial allele a frequency, we shall accept according to [3] $q_{f0} = 0.605$, see also formula (2.8).

The estimation of fitness $(1-s)$ is enough challenge. In paragraph 2.4 the analysis of this parameter has been submitted. It is shown that proceeding from some supervision the most comprehensible needs size is $1-s=0.29$. Therefore, for calculation we use the size of the selection parameter $s=0.71$.

In Figure 3.4 (curve 2, the right scale of the ordinates axis) the graph of dependence of the decrease in blood of the allele O frequency q_f in women on dimensionless time n at the selection plotted under the formula (3.39) is shown.

A decrease in blood of the alleles O frequency occurs on account of the elimination of mutant alleles. This process is rather intensive; at least, it is a much more intensively mutational process in hemophilia (Figure 3.4; curve 1, the left scale of the ordinates axis). Thereof concerning Haldane and Danforth [4, 5]'s ideas, some words are necessary to explain about the balance of mutation frequencies and selection in hemophilia. Such equilibration is hardly feasible. Physiologically, it involves two different processes. The law of mutations' occurrence (3.29) is absolutely not similar to the law of alleles' elimination in a population at selection (3.39). But actually equilibrations are not necessary either. It is only necessary that the velocity of mutations' occurrence does not exceed the velocity of selection.

3.5. Induced Mutagenesis in a Population

Mutagenesis in a population as well as in a family tree can be caused by various external factors. Among them it is necessary to allocate, first of all, chemical influence and radiating influence on a genome. Between these two kinds of influences there is no absolute border. At a radioactive irradiation of an organism there is the set of products of the water's radiolysis, including various radicals, active forms of oxygen which can cause chemical mutagen changes in molecules DNA. As the basic part of radicals recombine, not reaching a nucleus, the radicals formed near to a nucleus $\sim 3\,nm$ are most dangerous. In Table 7 the radiating damages arising in DNA of a cell at the doze 1 *gray* are shown [18].

Table 7.

Types of damage	Number of damages in a nucleus on 1 *gray*
One-filamentary breaks in DNA	1000
Damage to the bases in DNA	500
Bond DNA-albumen	150
Two-filamentary breaks in DNA	40

Only the small part of the DNA radiating damages is kept for a long time. The most part is eliminated in the course of DNA before replicative reparations. The cell has an enzymatic system (for example, endonuclease) cutting out the damaged sites and replacing them with normal nucleotides.

After irradiation some cells are carriers of so-called radiation - induced genome instability [18]. Thus, to the parts of cells, descendants are transferred an inclination to spontaneous formation of a chromosome aberrations *de novo* to the part of cells; descendants are transferred conformational changes in the structure of a chromatin. All of this results at a cellular level in a reduction in the reproductive ability of cells.

The nucleus of a cell can be directly subjected to the influence of radioactive particles: γ - quantums, α - and β - particles, proton and neutron radiation.

The probability *n* of hits of radioactive particles in a nucleus of the cell, owing to the fact that there can be genetic mutations, is described by the Poisson formula:

$$P = \frac{\alpha^n e^{-\alpha}}{n!}, \tag{3.40}$$

where $\alpha = S\Phi$ is the average number of hits in the nucleus for equal time intervals, S - the area of a nucleus section, Φ - so-called fluence - an average of the particles falling in a unit area, i.e., $\alpha = \Phi$ at $S = 1$ [18].

We shall assume that as a result of a hit in a nucleus, even one particle, some mutation of a gene is observed.

The probability of missing of a particle in the nucleus ($n = 0$), i.e., the probability of a bioobject survival, equals:

$$P_{sur} = \frac{(S\Phi)^0 e^{-S\Phi}}{0!} = e^{-S\Phi}. \tag{3.41}$$

Let the doze of irradiation D be such that $S\Phi = 1$. In this case the probability of a survival is $P_{sur} = e^{-1} = 0.37$, i.e., 37% of bioobjects survive. This doze of irradiation we designate D_{37}. The fluence Φ_{37} corresponds to this doze of the irradiation.

Taking into account $S\Phi_{37} = 1$ we find $S = \dfrac{1}{\Phi_{37}}$.

It is established that $D \sim \Phi$ [18], hence $\dfrac{D}{D_{37}} = \dfrac{\Phi}{\Phi_{37}} = S\Phi$.

Therefore, according to (3.41), the probability of a bioobject survival equals:

$$P_{sur} = e^{-\frac{D}{D_{37}}}. \tag{3.42}$$

This dependence refers to a «doze - effect».

In conditions of a real experiment the exponential dependence «doze - effect» is observed in the case of radiating inactivation by γ - rays of a viruses, enzymes, bacteria, spore, some unicells, i.e., relatively simple bioobjects.

3.5.1. Radiating Influence on a Panmictic Population

In paragraph 2.5.1 the question about the influence on a family tree of a long-acting, gradually decreasing radiation, owing to the radioactive decay of elements, has been considered.

We research a similar influence of radiation on a population. As the first approximation we shall consider the influence of radiation on a panmictic population.

It is simplified in that we assume that in the field of vital activity of a population only one radioactive element is scattered.

We believe as well as in paragraph 2.5.1 the capacity of a radioactive radiation doze is proportional to the activity of the radioactive elements scattered in the environment. For the frequency of mutations, taking into account the basic law of radioactive decay, we used the ratio $\mu = \frac{k}{2}\left(\frac{1}{2}\right)^{\left(\frac{T}{T_{1/2}}\right)n}$ where in this case it is designated that $T \sim 25\,years$ is the average time of one generation's alternation, $T_{1/2}$ - a half-life period of the radioactive element acting on a population, $n = \frac{t}{T}$ - dimensionless (normalized) current time of a population existence in conditions of radiation, k - some factor of proportionality. The factor 1/2 is used for the convenience of further calculations.

Let us note also that the activity of radioactive elements (the number of the disintegrated nucleus in the unit of time) is inversely proportional to their half-life period [8].

For the analysis of the variation of the recessive allele frequency q_f in the X-chromosome in women under the action of radiation it is possible to use the Hardy-Weinberg law for panmictic populations (3.18):

$$\frac{d^2 q_f}{dn^2} + \ln 2 \frac{dq_f}{dn} = \alpha , \qquad (3.43)$$

where α characterizes some mutagen factor of a population induced by the radiation.

Using the interrelationship between the mutagen factor and the frequency of the mutations $\alpha = 2\mu \ln 2$ found in paragraph 3.2.2, we find

$$\alpha = k \ln 2\left(\frac{1}{2}\right)^{\left(\frac{T}{T_{1/2}}\right)n} = k \ln 2R^n$$

where, as well as in paragraph 2.5.1, $R = \left(\frac{1}{2}\right)^{\left(\frac{T}{T_{1/2}}\right)}$ is used.

In the given designations the equation (3.43) will be transformed to the kind:

$$\frac{d^2 q_f}{dn^2} + \ln 2 \frac{dq_f}{dn} = k \ln 2R^n . \tag{3.44}$$

The general solving of the homogeneous equation:

$$\frac{d^2 q_f}{dn^2} + \ln 2 \frac{dq_f}{dn} = 0 \tag{3.45}$$

looks like:

$$q_{f\text{gen}} = C_1 + C_2 e^{-\ln 2n}, \tag{3.46}$$

where C_1 and C_2 are constants of integration.

We shall find the particular solution to the equation (3.44) as $q_f^* = LR^n$ where L is the unknown parameter. Substituting this solution in (3.44), we shall find:

$$LR^n \ln^2 R + \ln 2LR^n \ln R = k \ln 2R^n . \tag{3.47}$$

Hence, $L = \frac{k \ln 2}{\ln R \ln(2R)}$. Thus, the particular solution of the equation (3.44) looks like:

$$q_f^* = \frac{kR^n \ln 2}{\ln R \ln(2R)}. \qquad (3.48)$$

Writing down the solution of the equation (3.44) as the sum of the general solution (3.46) and the particular solution (3.48), we shall find the dependence of the intact allele frequency on time:

$$q_f = q_{f\text{gen}} + q_f^* = C_1 + C_2 e^{-\ln 2n} + \frac{kR^n \ln 2}{\ln R \ln(2R)} = \\ = C_1 + \left(C_2 + \frac{k(2R)^n \ln 2}{\ln R \ln(2R)} \right) e^{-\ln 2n}. \qquad (3.49)$$

We shall find constants C_1 and C_2 from conditions: at $n = 0$ the frequency $q_f = q_{f0}$ and initial speed of the frequency variation $\frac{dq_f}{dn} = 0$

As a result, we shall find:

$$C_1 = q_{f0} - \frac{k}{\ln R} \text{ and } C_2 = \frac{k}{\ln(2R)}. \qquad (3.50)$$

Hence, (3.49) will be transformed to the kind:

$$q_f = q_{f0} - \frac{k}{\ln R} + \frac{k}{\ln(2R)} \left(1 + \frac{(2R)^n \ln 2}{\ln R} \right) e^{-\ln 2n}. \qquad (3.51)$$

For a long duration of the influence of radiation on a population $n \to \infty$ the frequency of intact alleles after the action of radiation aspires to equilibrium $q_f \to q_{f\text{eq}}$ since $R < 1$. According to (3.51), the equilibrium frequency equals:

$$q_{feq} = q_{f0} - \frac{k}{\ln R}. \qquad (3.52)$$

If $q_{f0} < \frac{k}{\ln R}$, the equilibrium frequency is not reached and the damaging allele is eliminated from a population.

Let us note that the equilibrium frequency q_{feq} of a non-defective allele in a population does not depend on the period of a radioactive element's half-life. The essential difference in the influence on a population consists in the influence of radiating conditions on a family tree.

Taking into account $k \sim -\frac{N_0}{T_{1/2}}$, where N_0, there is a general initial quantity of the radioactive substance acting on a population (see paragraph 2.5.1), and $R = \left(\frac{1}{2}\right)^{\left(\frac{T}{T_{1/2}}\right)}$; we shall find

$$\frac{k}{\ln R} \sim -\frac{N_0}{T_{1/2}\frac{T}{T_{1/2}}\ln\left(\frac{1}{2}\right)} = \frac{1.443 N_0}{T} \neq f\left(T_{1/2}\right).$$

For calculation we shall accept $k = -\frac{W}{T_{1/2}}$. Parameter W is proportional to the initial quantity of the radioactive substance acting on a population $W \sim N_0$.

In Figure 3.5 the graphs of dependences plotted under the formula (3.51) frequencies q_f of non-defective alleles in women on dimensionless time $n = \frac{t}{T}$ are shown at various values of parameter W and the half-life period $T_{1/2} = 100\ years$ of radioactive substance scattered in a zone of a population's vital activity.

The parameter $W = 50$ corresponds to calculations of paragraph 2.5.1 and 2.5.2.

From the figure it can be seen that at $W = 16$ and $W = 50$ the equilibrium frequency of the intact allele is not established. It is connected to plenty of radioactive substances scattered in a zone of a population-vital activity.

In spite of the fact that the equilibrium frequency of the intact allele does not depend on a half-life period of radioactive substance, the character of the frequency q_f of the non-defective allele approach to the balance for various half-life periods is various.

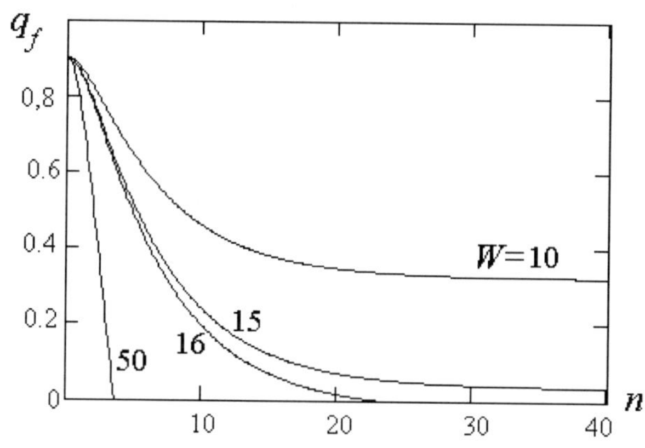

Figure 3.5.

In Figure 3.6 the dependence of the intact allele frequency q_f after the action of radiation in women on dimensionless time $n = \frac{t}{T}$ is shown at various values of a half-life period $T_{1/2}$ of radioactive substance for parameter $W = 15$.

In Figure 3.6 it can be seen that at small half-life periods and, accordingly, the big activity of radioactive substance [8] there is a faster elimination of the defective allele from a population.

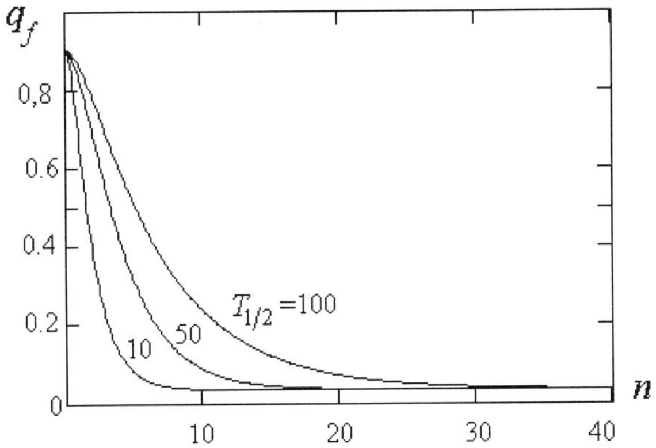

Figure 3.6.

3.5.2. Incomplete Inheritance in a Population

In paragraph 2.6 the question about the incomplete inheritance of recessive alleles to the man from the mother in a family tree, where the genetic reasons for this phenomenon have been specified, has been considered. We will now research incomplete inheritance in a population.

For this purpose we shall transform the formula (2.11) $q_{f(n+1)} = \frac{1}{2}\left(q_{fn} + q_{mn}\right)$ for finding the recessive allele frequency of X-chromosomes in the following generation of women to a kind:

$$q_{fn} - 2q_{f(n+1)} + q_{mn} = 0. \tag{3.53}$$

Similarly to the analysis of a family tree in paragraph 2.6, we shall use the parameter of incomplete inheritance α which shows the part of recessive alleles which the man does not inherit from his mother $q_{mn} = (1-\alpha)q_{f(n-1)}$.

In this case the formula (3.53) can be copied as:

$$q_{fn} - 2q_{f(n+1)} + (1-\alpha)q_{f(n-1)} = 0. \tag{3.54}$$

We carry out (similarly to paragraph 3.1) the transition from a family tree to a population.

We shall write down the differential equation for the recessive allele frequency q_{fn} in women in a population in a general form:

$$\frac{d^2 q_{fn}}{dn^2} + \eta \frac{dq_{fn}}{dn} = \mu q_{fn}, \tag{3.55}$$

where η and μ are constants.

Let us transform (3.55) to finite-differential form:

$$\frac{q_{f(n+1)} - 2q_{fn} + q_{f(n-1)}}{\Delta n^2} + \eta \frac{q_{fn} - q_{f(n-1)}}{\Delta n} = \mu q_{fn}. \tag{3.56}$$

Uniting similar members in (3.56) and multiplying the result by -2 we shall find:

$$-2q_{f(n+1)} + 2\left(2 - \eta\Delta n + \mu\Delta n^2\right)q_{fn} - 2(1 - \eta\Delta n)q_{f(n-1)} = 0. \tag{3.57}$$

Let us identify formulas (3.57) and (3.54). As a result we have a system of two algebraic equations:

$$2\left(2 - \eta\Delta n + \mu\Delta n^2\right) = 1, \tag{3.58}$$

$$-2(1 - \eta\Delta n) = 1 - \alpha. \tag{3.59}$$

The solution to this system looks like:

$$\eta = \frac{3 - \alpha}{2\Delta n}. \tag{3.60}$$

$$\mu = -\frac{\alpha}{2\Delta n^2}. \tag{3.61}$$

Hence, the differential equation (3.55) can be found in the following kind:

$$\frac{d^2 q_f}{dn^2} + \left(\frac{3-\alpha}{2\Delta n}\right)\frac{dq_f}{dn} + \frac{\alpha}{2\Delta n^2} q_f = 0. \tag{3.62}$$

The number of the generation does not play a role in the population of a continuous time scale; therefore, the index n at alleles frequency is usually omitted.

The found linear differential equation (3.62) is the Hardy-Weinberg law for a population with accounts for the recessive allele incomplete inheritance to a man from his mother. At $\alpha = 0$ (complete inheritance) the equation (3.62) will be transformed to the kind:

$$\frac{d^2 q_f}{dn^2} + \frac{3}{2\Delta n}\frac{dq_f}{dn} = 0, \tag{3.63}$$

which corresponds to the equilibrium panmictic population; see (3.7).

Let us write down the differential equation (3.62) as the equation of free (damped) oscillations [19]:

$$\frac{d^2 q_f}{dn^2} + 2\beta\frac{dq_f}{dn} + \omega_0^2 q_f = 0, \tag{3.64}$$

where $2\beta = \frac{3-\alpha}{2\Delta n}$ is designated the double factor of oscillations attenuation of the recessive alleles frequency and $\omega_0^2 = \frac{\alpha}{2\Delta n^2}$ - the square of natural cyclic frequency of these oscillations.

The equation (3.64) is solved analytically [19]:

$$q_f = q_{f\,\max} e^{-\beta n} \sin(\omega n + \varphi), \tag{3.65}$$

where $q_{f\max}$ is the peak value of the frequency of the recessive alleles in women for population-genetic oscillations, and φ - is the initial phase of oscillations.

The cyclic frequency of oscillations equals [19]:

$$\omega = \sqrt{\omega_0^2 - \beta^2} = \sqrt{\frac{\alpha}{2\Delta n^2} - \left(\frac{3-\alpha}{4\Delta n}\right)^2} = \frac{1}{4\Delta n}\sqrt{8\alpha - (3-\alpha)^2} = \qquad (3.66)$$
$$= \frac{1}{4\Delta n}\sqrt{8\alpha - 9 + 6\alpha - \alpha^2} = \frac{1}{4\Delta n}\sqrt{-9 + 14\alpha - \alpha^2}$$

Positive (or zero) value under the root in view of the top border of the incomplete inheritance α parameter takes place at $0.68 \le \alpha \le 1$. Thus, formally there is some range of sizes α at which there can be population-genetic oscillations. The natural cyclic frequency of oscillations $\omega_0 = \frac{1}{\Delta n}\sqrt{\frac{\alpha}{2}}$ is in a range $\frac{0.58}{\Delta n} \le \omega_0 \le \frac{0.71}{\Delta n}$. Using $\Delta n = \frac{3}{2\ln 2} \approx 2.16$ (see paragraph 3.1), we shall estimate this range $0.27 \le \omega_0 \le 0.33$.

The periods of oscillations $T_0 = \frac{2\pi}{\omega_0}$ corresponding to these natural frequencies is in the limits $17 \le T_0 \le 23$. Taking into account the time of one generation being 25 years, the period of natural oscillations of the recessive allele frequency is in the range $425\ years \le T_0 \le 575\ years$.

An essential limitation in using the solution (3.65) is positivity of the function q_f.

Let us consider two particular solutions of the equation (3.64) at $q_{f\max} = 1$:

$$q_f = e^{-\beta n}\sin(\omega n), \qquad (3.67)$$

$$q_f = e^{-\beta n}\cos(\omega n). \tag{3.68}$$

If we were to examine a pendulum as a mechanical analogy of free oscillations, the formula (3.67) is realized in the case of an external sharp push of a pendulum from a position of balance, and the formula (3.68) at an initial external deviation of a pendulum from a position of balance.

In Figure 3.7 the graphs of dependence of the recessive allele frequency in a population on time are shown at various parameters of the incomplete inheritance plotted under formulas (3.67) at $\alpha = 0.99$ and (3.68) at $\alpha = 0.69$.

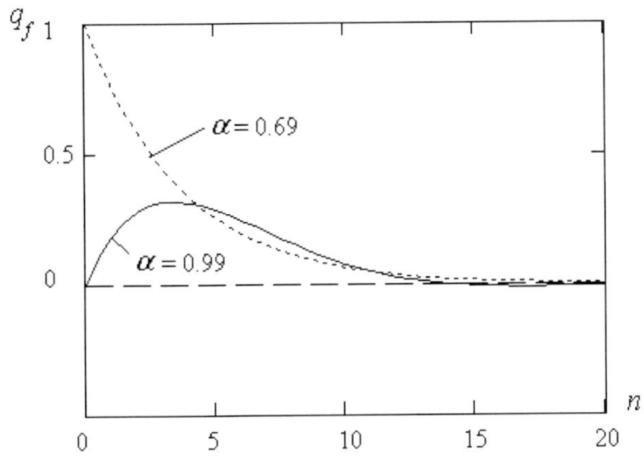

Figure 3.7.

In connection with the large size of the attenuation factor (for $\alpha = 0.99$ the size $\beta = \dfrac{3-\alpha}{4\Delta n} = 0.23$ and for $\alpha = 0.69$ the size $\beta = 0.27$) the oscillations of full value do not arise. More positive values of function q_f in the formula (3.67) are observed at great values of the incomplete inheritance parameter α than in the formula (3.68).

Due to regulation of the size of the incomplete inheritance parameter α ($1-\alpha$ proportional to number of Okazaki inherited fragments) can be realized as the process of initial growth of the recessive allele frequency q_f (the formula (3.67)) or process of initial reduction of this frequency (the formula (3.68)).

Thus, depending what is realized from formulas (3.67) or (3.68), first, the corresponding size of the incomplete inheritance parameter α (big or small) should be used; secondly, the frequency q_f of the recessive allele can grow or fall from the initial generation. In nature, apparently due to regulation of the incomplete inheritance parameter, the law of change of the recessive allele frequency (3.67) or (3.68) is realized.

Let us note that in nature if any process is allowed by the law (in our case, Hardy-Weinberg genetic law) this process is usually realized for any purposes.

A feature of the performed analysis of the Hardy-Weinberg law for a population will be that this law at incomplete inheritance allows for a variation of the recessive allele frequency, with both increases and reduction. The considered processes are not connected to natural selection, as for example, with inbreeding (see paragraph 3.6.1), i.e., with the Darwinian type of evolution. Therefore, it is possible to admit to the probability of the existence of evolution, not the Darwinian type of population evolution. This type of evolution can be included due to the incomplete inheritance of the *X*-chromosome by men if the adaptation of a population to changed external conditions is necessary. Thus, all changes occur quickly enough in generations $n < 5$ or in a period of $t < 125$ years.

For example, for a long time it was noticed that before the war more boys were born. This phenomenon can be explained only with the help of a non-Darwinian type of evolution when the psychological pressure upon a population includes the genetic mechanism of incomplete inheritance. Another example is the ideological pressure placed upon enterprising and hardworking people in some countries which has led to non-Darwinian evolution which in generations ceases to reproduce a similar type of people.

3.6. NONLINEAR GENETICS

In paragraph 3.1 it is shown that variation of the recessive (or dominant) genes linked to the *X*-chromosome frequency during the time of a population life in condition of the balance of a population is described by the linear differential equation of the second order (3.15) or (3.15a). This equation expresses the Hardy-Weinberg law for a panmictic population. Also, the linear differential equations of the second order describe, for example, the mutational process resulting in a hemophilia (see paragraph 3.2.2), or the mutational process under action of radiation (see paragraph 3.5.1), etc. Similar processes can be attributed to processes of linear genetics.

However, in genetics there are processes resulting in nonlinear differential equations. These processes concern nonlinear genetics. As an example of a similar process, we shall consider the development of a population in the presence of inbreeding.

3.6.1. Inbreeding in a Population

The problem of the inbreeding influence on the population is the key problem of genetics.

For a family tree the problem of inbreeding has been investigated in detail in paragraph 2.7.

In paragraph 2.7.1 it has been shown that in inbred family trees and populations there is an accumulation called the genetic load arising basically due to growth in the recessive homozygotes.

Let us carry out the analysis of the inbreeding action on a population.

For research on inbreeding in populations we shall not take into account the action of selection (it will be taken into account later) to avoid complicating the analysis. This is allowable since selection submits to other laws rather than the occurrence of mutations owing to inbreeding.

Using a standard technique of transition from a family tree to a population in the basis of the analysis we shall use the formula (2.85) which allows us to calculate the fraction of the recessive allele a in the X-chromosome of women in the following generation as the sum of the equilibrium term, gradually increasing the genetic load owing to inbreeding.

Taking into account that the man inherits the X-chromosome from his mother i.e., $q_{mn} = q_{f(n-1)}$, it is possible to transform (2.85) thus:

$$q_{fn} - 2q_{f(n+1)} + q_{f(n-1)} = -2\frac{F}{1-F} q_{f(n-1)} q_{fn}, \tag{3.69}$$

where F is the inbreeding factor for a population (see paragraph 2.7.1).

The differential equation for the change of frequency q_{fn} of recessive allele a in the inbred population shall be written down in the general form:

$$\frac{d^2 q_{fn}}{dn^2} + \eta \frac{dq_{fn}}{dn} = \mu q_{fn}^2, \tag{3.70}$$

where the independent variable n is in this case the time of the population life normalized on average in family trees of the population time of one alternation of generation (approximately 25 years) i.e., actually continuous dimensionless time, η and μ - constants.

Let us pass in (3.70) to finite-differential form:

$$\frac{q_{f(n+1)} - 2q_{fn} + q_{f(n-1)}}{\Delta n^2} + \eta \frac{q_{fn} - q_{f(n-1)}}{\Delta n} = \mu q_{fn}^2. \tag{3.71}$$

Let us collect the like terms:

$$q_{f(n+1)} - (2 - \eta\Delta n) q_{fn} + (1 - \eta\Delta n) q_{f(n-1)} = \mu\Delta n^2 q_{fn}^2. \tag{3.72}$$

Let us multiply (3.72) by -2; in addition, we shall replace $q_{fn}^2 = q_{f(n-1)} q_{fn}$ which is allowable for the continuous time of the generation alternation scale i.e., for a population:

$$-2q_{f(n+1)} + 2(2 - \eta\Delta n) q_{fn} - 2(1 - \eta\Delta n) q_{f(n-1)} = -2\mu\Delta n^2 q_{f(n-1)} q_{fn} \tag{3.73}$$

Let us identify (3.73) and (3.69). As a result, we have a system of three algebraic equations:

$$2(2 - \eta\Delta n) = 1, \tag{3.74}$$

$$-2(1 - \eta\Delta n) = 1, \tag{3.75}$$

$$\mu\Delta n^2 = \frac{F}{1 - F}. \tag{3.76}$$

The solution of the given system looks like:

$$\eta = \frac{3}{2\Delta n}, \tag{3.77}$$

which is similar to the equilibrium of the panmictic population; see formula (3.6).

$$\mu = \frac{F}{\Delta n^2 (1-F)}. \tag{3.78}$$

Hence, the differential equation (3.70) can be copied as:

$$\frac{d^2 q_f}{dn^2} + \frac{3}{2\Delta n}\frac{dq_f}{dn} = \frac{F}{\Delta n^2 (1-F)} q_f^2. \tag{3.79}$$

The number of the generation for the population does not play a role – therefore, for the continuous time scale an index n is omitted.

The found nonlinear differential equation (3.79) is the Hardy-Weinberg law for a population with the account of inbreeding.

At $F = 0$ (inbreeding is absent) the equation (3.79) will be transformed to the kind that corresponds to the equilibrium of the panmictic population:

We use in (3.79) the new independent variable $t = \frac{n}{\Delta n}\sqrt{\frac{F}{1-F}}$.

If we are to accept (see paragraph 3.1) $\Delta n = \frac{3}{2 \ln 2}$, we shall find

$$t = \frac{2 \ln 2n}{3}\sqrt{\frac{F}{1-F}} \approx 0.462n\sqrt{\frac{F}{1-F}}.$$

For the new variable t the equation (3.79) depends only on one constant parameter ε :

$$\frac{d^2 q_f}{dt^2} + \varepsilon \frac{dq_f}{dt} = q_f^2, \tag{3.80}$$

where it is designated $\varepsilon = \frac{3}{2}\sqrt{\frac{1-F}{F}}$. At the variation of the inbreeding factor $0 \le F \le 1$ the size is $\infty \ge \varepsilon \ge 0$.

The nonlinear differential equation (3.80) can be solved only numerically.

Initial conditions first of all are necessary for the solution of the differential equation (3.80): initial frequency q_{f0} of the recessive allele *a* and initial speed of increase in this frequency $\left(\frac{dq_f}{dt}\right)_0$ or $\left(\frac{dq_f}{dn}\right)_0$.

Initial frequency q_{f0} depends on the time of the beginning of the population analysis. Therefore, without reduction of the analysis generality at $t=0$ or $n=0$, we shall accept $q_{f0}=0.6$.

Various variants of numerical calculations under the formula (3.80) show that the result of the calculation at a small initial speed of the frequency increase q_f practically does not depend on this speed. Therefore, $\left(\frac{dq_f}{dt}\right)_0 = \left(\frac{dq_f}{dn}\right)_0 = 0$ was accepted. The step of the variable t equals $\Delta t = 0.001$ and numbers of steps 20000 at calculation were used.

In Figure 3.8 the calculation of the population dynamics of recessive allele *a* on the basis of the equation (3.80) at various inbreeding factors *F* is shown. On the abscissa axis the size $n \approx \frac{t}{0.462}\sqrt{\frac{1-F}{F}}$ is plotted

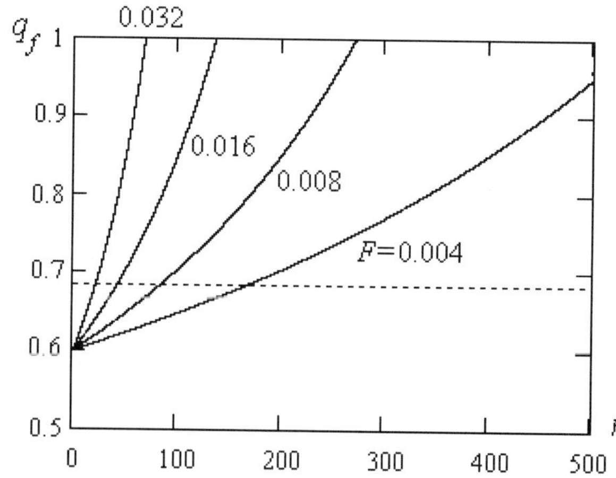

Figure 3.8.

From the graphs it can be seen that at the presence of inbreeding in the population there is a continuous increase of the recessive allele *a* frequency q_f in which lethal adverse (at association in homozygotes) mutations can arise i.e., there is a negative genetic load in the population. The growth of the recessive allele frequency received in calculation is observed due to the growth in the frequency of the recessive homozygotes *aa*. Thus, the frequency of heterozygotes *Aa* at inbreeding falls, which leads to a general decline in the intact recessive allele *a* frequency q_f on all zygotes in a population (*AA*, *Aa*, *aa*) due to adverse mutations only in the recessive homozygotes *aa*.

Traditionally [21] the genetic load is estimated on the basis of the analysis of deadborns or children's death rate with the help of two parameters *A* and *B* included in the approximate equation for the fraction of the survived zygotes:

$$S = \exp(A + BF). \tag{3.81}$$

Unfortunately, the data on the experimental sizes of parameters *A* and *B* have a great variation in values. In [3] the data on the population of France $B = 1.5 - 2.5$ and $\frac{B}{A} = 15.06 - 21.41$ is cited.

More detailed data are available on the population of Japan. In different prefectures the ratio $\frac{B}{A} = 15.2 - (-5)$. Negative value $\frac{B}{A}$ means that the children's death rate in consanguineous marriages was lower than in non-consanguineous marriages [3]. Average sizes in the country are $A = 0.1036$, $B = 0.67$ and $\frac{B}{A} = 6.7$.

It is possible to assume [3] that there is some threshold of a recessive homozygous genes allowable frequency compatible with the survival of an individual.

Taking into account values *A* and *B*, and also the inbreeding factor for Japan $F = 0.004$ [3], under the formula (3.81) we shall find the fraction of the survived zygotes $S = 0.9$. The fraction of the perish zygotes there is $1 - S = 0.1$. The elementary estimation under the formula $1 - S = q_f^2 = 0.1$ gives the fraction of lethal recessive alleles at their association in homozygous

$q_f \approx 0.316$ [7]. Hence, the proportion of non-mutant genes equal to the threshold frequency of the recessive alleles is $q_{f\,max} = 1 - q_f = 0.684$.

In Figure 3.8 the dotted line shows the threshold frequency. For Japan this frequency is attained approximately in generation 167.

The received result shows that the accumulation of the genetic load is a very slow process. Achievement of the threshold frequency for Japan to be carried out only in 167 generations or 4175 years with the time of one generation alternation given as $T = 25$ years.

In [3] it is noticed that from the beginning of the population development with inbreeding factor $F = 0.003 - 0.005$ up to the achievement of an equilibrium frequency which is established at the action of selection approximately 4500 years should elapse. In calculation the formula of the linear increase of the frequency of recessive alleles was used at mutation speed $\mu = 10^{-5}$. The increase of the mutant alleles number was compensated by selection against recessive homozygotes with parameter of selection 0.5.

Therefore, it is most likely that there are limiting values of the genetic load in a human population of many countries, especially taking into account the small sizes of inbreeding factor (Canada $F = 45 \cdot 10^{-5}$, France $F = 23 \cdot 10^{-5}$, Germany $F = 19 \cdot 10^{-5}$, Italy $F = 16 \cdot 10^{-5}$ [3] etc.) even at the presence of selection until now have not been achieved.

In our opinion, with the help of the offered method it is possible to examine not only the genetic load related with inbreeding that is harmful for the population but also useful and indifferent genetic loads (or attributes).

In the early stages of mankind's development the role of inbreeding was more in connection with a smaller number of people and the isolation of separate groups. Therefore, the accumulation of the genetic load was more intensive (the inbreeding factor was much greater). After the exodus from Africa and moving to northern Europe, primeval people began to live in conditions of smaller light exposure including in the ultra-violet part of spectrum. As a result, those with skin of originally dark pigmentation started to have a whiter color. It promoted the occurrence of vitamin D, the deficiency of which results in rickets [3]. The process of dark pigmentation reduction in the skin can be examined as an accumulation of the recessive useful genetic load. Natural selection fixed a given phenotypic attribute in the people of the region.

To the indifferent genetic load in a population it is possible to ascribe an occurrence such as a recessive attribute like blond hair, for example, in the northern part of Europe where apparently in the early stages the inbreeding was especially widespread. The color of hair does not play a role in the survival of a population. The fixed nature of this attribute was carried out by the sexual selection i.e., marriage preferences of people in the region.

However, it is traditionally accepted that the accumulation of the genetic load in the population is determined by harmful mutations. Some fears of H. J. Muller are connected to this process about the danger of the future biological degeneration of mankind [12].

3.6.2. Why It Was Not Necessary to Kill Giraffe Marius in the Copenhagen Zoo

Let us consider one excess directly connected to an inbreeding problem. It is the murder of the giraffe Marius in the Copenhagen Zoo on February, 9, 2014, which was explained as a necessity to avoid inbreeding in a population.

I want to note immediately that while writing this the author does not have an emotional attitude to the fact of an animal murder. The author believes that the murder of animals is quite allowable; first, for obtaining meat food-stuffs, and, second, for medical and biological research.

Let us apply the analysis of an inbreeding problem carried out earlier, i.e., consanguineous mating, to a concrete case in the Copenhagen Zoo.

We shall begin by defining that the giraffe Marius was connected to the separate family tree in which there was already some genetic load which had been saved up throughout its existence by this family tree both in the wild and in a zoo.

We shall lead the analysis in three directions: we shall take into account the influence of inbreeding in a family tree on autosomal genes; we shall take into account the influence of inbreeding in a family tree on sex-linked genes; and also, we shall take into account the influence of inbreeding on a population where the accumulation of a negative genetic load is possible.

In paragraph 2.7 it was shown that the separate act of a consanguineous mating has only small importance for a family tree.

The formula (2.75) specifies there is no increase in frequency of any allele in autosomes in the posterity of the inbred individuals. Therefore, there is equally no increase in the probability of the genic mutations in the posterity of this individual.

A similar conclusion can be made for genes in sex-linked chromosomes. The formula (2.81) shows in the following generation there is a restoration of a panmictic approach to the balance. The probability of genic mutations in the posterity of inbred individuals does not change in comparison with a panmictic family tree. Thus, and for the genes linked to the X-chromosome as well as for genes of autosomes, the separate act of a consanguineous mating has little importance for a family tree.

Therefore, from the point of view of the giraffe Marius, the influence of consanguineous mating on an autosomal genome and on a genome linked to the X-chromosome his exception from a family tree does not lead to any positive consequences.

Let us pass to the analysis of a genetic load in populations of giraffes.

Can a giraffe from a zoo with a raised level of genetic load influence a wild population of African giraffes after being liberated? We shall consider this question in more detail.

It is possible to assume that in populations of African giraffes the inbreeding factors are much higher than in humans where there are ethical and religious interdictions on consanguineous mating. This makes the killing of the giraffe Marius even more senseless. Populations of African giraffes have existed for thousands of years and have not yet faced extinction due to the consequences of inbreeding. An accumulation of a genetic load in populations is, as shown in the previous analysis (see paragraph 3.6.1), a very slow process (Figure 3.8), taking place over hundreds of generations. The additional negative mutations which have arisen owing to an increase of the recessive allele frequency in homozygotes are eliminated from a population due to natural selection. This slows down or excludes the accumulation of a genetic load in a population.

In a zoo an animal can arrive from the wild with that or other genetic load – a consequence of a population's enduring existence. Ways of defining a genetic load level do not currently exist. In a zoo there should be consanguineous mating during lines of generations for the further accelerated accumulation of a genetic load. And the intensity of inbreeding in a zoo should exceed the intensity of inbreeding in the wild. Even if we are to assume that such conditions exist, upon the liberation of an animal from the zoo into the wild, the animal and his nearest posterity from a family tree with a raised level of genetic load would soon be eliminated from a population due to natural selection. Therefore, there is no advantage in the complete exclusion of inbreeding in the family tree of zoo animals from a genetic point of view.

There cannot be any harm done to a population of giraffes as a result of the individual act of the giraffe Marius consanguineously mating.

The purpose of a population geneticist is not the prevention of individual acts of consanguineous mating in a family tree of zoo animals or efforts to protect wild populations from the pressure of a human civilization. Thus, there should not be any intervention in the inbred processes of a population. Thus, the murder of the giraffe Marius in the Copenhagen Zoo was absolutely not necessary from a genetic point of view.

3.6.3. Radiating Influence on an Inbred Population

We will now research the influence of radiation on an inbred population.

As a basis for the analysis we shall use the differential equation (3.44) determining the influence of radiation on a panmictic population having added similarly (3.79) in the right part nonlinear term determining inbreeding. As a result, we shall receive:

$$\frac{d^2 q_f}{dn^2} + \frac{3}{2\Delta n}\frac{dq_f}{dn} = \frac{F}{\Delta n^2 (1-F)} q_f^2 + k \ln 2R^n . \quad (3.82)$$

The equation (3.82) has the analytical particular solution. We shall find it. We shall assume that the right part (3.82) is equal to zero:

$$\frac{F}{\Delta n^2 (1-F)} q_f^2 + k \ln 2R^n = 0 . \quad (3.83)$$

From the given condition it is possible to find the recessive allele frequency:

$$q_f = \Delta n R^{\frac{n}{2}} \sqrt{-\frac{(1-F)}{F} k \ln 2} . \quad (3.84)$$

Substituting (3.84) in the formula (3.82), we shall find:

$$\Delta n R^{\frac{n}{2}} \frac{\ln^2 R}{4} \sqrt{-\frac{(1-F)}{F} k \ln 2} + \Delta n R^{\frac{n}{2}} \frac{\ln R}{2} \sqrt{-\frac{(1-F)}{F} k \ln 2} \frac{3}{2\Delta n} = 0$$

(3.85)

The equation (3.85) is carried out under the condition of:

$$\Delta n = -\frac{3}{\ln R}. \tag{3.86}$$

Taking into account the interrelationship of parameter R with a half-life period $T_{1/2}$ of the radioactive substance working on a population $R = \left(\frac{1}{2}\right)^{\left(\frac{T}{T_{1/2}}\right)}$ see paragraph 2.5.1, where $T \approx 25\ years$ is the average time of a generation's alternation, we find $\Delta n = \frac{3T_{1/2}}{T \ln 2}$. It is interesting to note at a half-life period two times smaller than the average time of the generation's alternation $\frac{T}{T_{1/2}} = 2$ the size $\Delta n = \frac{3}{2 \ln 2}$ that corresponds to a panmictic population without external influence on it (see paragraph 3.1).

Unfortunately, the formula (3.84) is difficult to use for concrete calculations since the initial condition is rigidly set: at $n = 0$ the recessive allele frequency is $q_{f0} = -\frac{3}{\ln R} \sqrt{-\frac{(1-F)}{F} k \ln 2}$.

For the analysis of the equation (3.82) it is possible to see that the right part of this equation describes two opposite tendencies of the recessive allele frequency q_f variation. The square-law term $\sim q_f^2$ gives the tendency of this frequency growth (see the formula (3.79) and Figure 3.8). The term with exponential function $\sim R^n$ gives the falling dependence of the recessive allele frequency (see the formula (3.44) and Figure 3.5). Therefore, it is necessary to expect the existence of a function q_f minimum.

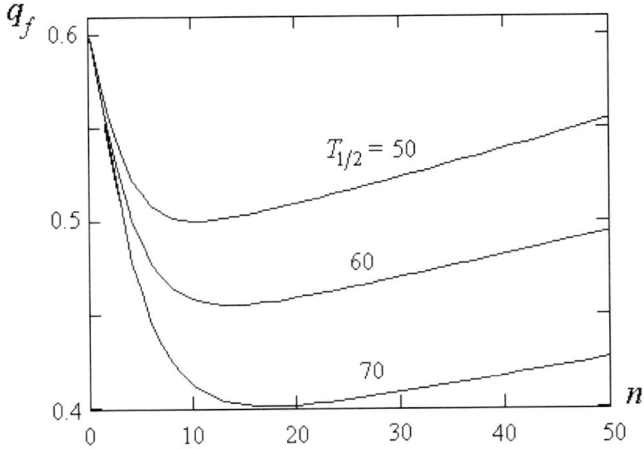

Figure 3.9.

In Figure 3.9 the result of the numerical solution of the equation (3.82) is shown at the various half-life periods ($T_{1/2} = 50$ *years* , 60 *years* and 70 *years*) of radioactive substance working on an inbred (inbreeding factor $F = 0.008$) population.

For calculation the following initial conditions were used: at $n = 0$ the allele frequency $q_{f0} = 0.6$, speed of change of the allele frequency $\left(\frac{dq_f}{dn}\right)_0 = 0$ as well as in the case of radiating influence absence, see paragraph 3.6.1.

Constant $k = -\frac{W}{T_{1/2}}$ is the parameter proportional to the initial quantity of radioactive substance. The parameter $W = 0.05$ was accepted as being very much smaller than in similar calculations of a panmictic population (see paragraph 3.5.1), with the purpose of balance concerning the weak influence of inbreeding and powerful mutagenic radiating influence; i.e., the quantity of radioactive substance in an environment was necessarily rather small. The average time of the generation's alternation is accepted as being $T = 25$ *years* .

In calculation sizes $\Delta n = \dfrac{3T_{1/2}}{T \ln 2} = 12.12$ and $R = \left(\dfrac{1}{2}\right)^{\left(\frac{T}{T_{1/2}}\right)} = 0.781$ were used.

In Figure 3.10 the result of the numerical solution of the equation (3.82) is shown at various inbreeding factors $F = 0.02, 0.015, 0.01$ and under condition of action on the inbred population of radioactive substance with the half-life period $T_{1/2} = 70\ years$. All other parameters of calculation were similar to the previous calculation (see Figure 3.9).

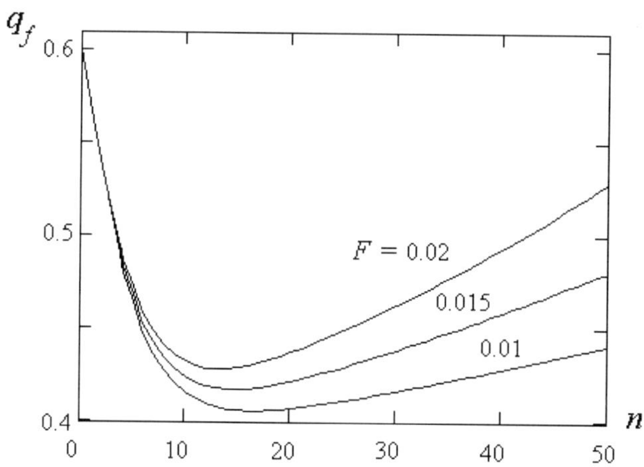

Figure 3.10.

As follows from Figure 3.9 and Figure 3.10, at the influence of a radioactive substance on an inbred population the existence of the minimal value of recessive allele frequency q_f is observed. This value at parameters used in calculations is attained in the interval $n = 10 \div 20$ or at the time interval $t = 250 \div 500\ years$.

Originally, the basic influence on a population is rendered with radiation. However, in the process of radioactive substance disintegration the inbreeding action leading to the growth of the recessive allele frequency starts to prevail.

At the influence of radiation on a population the frequency of healthy recessive alleles is reduced due to their mutations and is further eliminated from a population. Over time the inbreeding results in the growth of the healthy recessive alleles in homozygotes. The growth of the recessive allele frequency q_f for the account of inbreeding is not able to fill their mutation due to the influence of radiation. However, at some level of frequency q_f the probably full restoration of the recessive allele frequency will perish due to radiation only in the recessive homozygotes.

The increase in recessive homozygote frequency at inbreeding occurs due to the reduction of these alleles' frequency in heterozygotes. In the following generation in the presence of inbreeding the allele frequency is kept both in autosomes (2.75) and in sex-linked chromosomes (2.81). At the inbreeding the alleles are simply redistributed between homozygotes to more increases and heterozygotes to more reduction. Therefore, despite the increase in recessive allele frequency in homozygotes as a whole, the frequency of these alleles on all zygotes upon the mutagen influence of radiation decreases.

3.6.4. Natural Selection in a Population as a Problem of Nonlinear Genetics. Power of Selection

The problem of natural selection has been mentioned for a family tree in paragraph 2.4 and for a population in paragraph 3.4. The basic characteristics of selection have been introduced: parameter of selection s and fitness of a genotype $1-s$. However, in paragraph 3.4 the simplified technique of algebraic interrelation (3.39) reception between the blood recessive allele O frequency q_f in women in the case of hemophilia and dimensionless time n in the presence of selection has been used.

In the given paragraph we shall solve the problem of selection in a population using a standard method of transition from finite-differential equation characteristic for a family tree to the differential equation correct for a population. This way is submitted, for example, in paragraph 3.1.

As before, we shall carry out the analysis of natural selection by the example of selection against recessive homozygotes.

In paragraph 3.4 the recurrent ratio (3.34) was received for the finding of a recessive allele frequency in women if those frequencies in the two previous generations of women are known:

$$q_{fn} = \frac{\frac{1}{2}\left(q_{f(n-1)} + q_{f(n-2)}\right) - q_{f(n-2)} q_{f(n-1)} s}{1 - q_{f(n-2)} q_{f(n-1)} s}. \tag{3.87}$$

In the formula (3.87) s there is a standard parameter of selection against the recessive homozygotes.

Through simple algebraic transformations of the formula (3.87) we shall find:

$$q_{f(n-1)} + q_{f(n-2)} - 2q_{fn} = 2s\left(q_{f(n-2)} q_{f(n-1)} - q_{fn} q_{f(n-2)} q_{f(n-1)}\right) \tag{3.88}$$

Let us consider the differential equation:

$$\frac{d^2 q_{f(n-1)}}{dn^2} + \eta \frac{dq_{f(n-1)}}{dn} = \alpha s\left(q_{f(n-2)} q_{f(n-1)} - q_{fn} q_{f(n-2)} q_{f(n-1)}\right), \tag{3.89}$$

where η and α are constants. Differentiation in (3.89), as before, goes on dimensionless time n (normalized on average time of generation alternation in a population).

Let us pass in (3.89) to the finite-differential form of derivatives:

$$\begin{aligned} &\frac{q_{fn} - 2q_{f(n-1)} + q_{f(n-2)}}{\Delta n^2} + \eta \frac{q_{f(n-1)} - q_{f(n-2)}}{\Delta n} = \\ &= \alpha s\left(q_{f(n-2)} q_{f(n-1)} - q_{fn} q_{f(n-2)} q_{f(n-1)}\right) \end{aligned}. \tag{3.90}$$

Transforming (3.90) by analogy to paragraph 3.1, we shall find:

$$\begin{aligned} &-2q_{fn} + 2(2 - \eta\Delta n) q_{f(n-1)} - 2(1 - \eta\Delta n) q_{f(n-2)} = \\ &= -\alpha 2s\Delta n^2\left(q_{f(n-2)} q_{f(n-1)} - q_{fn} q_{f(n-2)} q_{f(n-1)}\right) \end{aligned}. \tag{3.91}$$

Let us identify the equation (3.91) with the equation (3.88). In paragraph 3.4 as a result of a similar identification the size $\eta = \frac{3}{2\Delta n}$ has been found, see the formula (3.6). Furthermore, it is possible to see that $\alpha = -\frac{1}{2\Delta n^2}$.

Hence, the differential equation (3.89) can be written down as:

$$\frac{d^2 q_{f(n-1)}}{dn^2} + \frac{3}{2\Delta n}\frac{dq_{f(n-1)}}{dn} = -\frac{1}{2\Delta n^2} s\left(q_{f(n-2)}q_{f(n-1)} - q_{fn}q_{f(n-2)}q_{f(n-1)}\right) \tag{3.92}$$

At the transition to a population the number of the generation does not play a role; therefore, we shall receive:

$$\frac{d^2 q_f}{dn^2} + \frac{3}{2\Delta n}\frac{dq_f}{dn} = -\frac{s}{2\Delta n^2}\left(q_f^2 - q_f^3\right). \tag{3.93}$$

The nonlinear differential equation (3.93) allows us to calculate the frequency q_f of intact recessive allele; for example, of blood O in women in the *X*-chromosomes of a population in the presence of selection on mutant alleles. For the convenience of calculation the new independent variable $t = \frac{n}{\Delta n}\sqrt{\frac{s}{2}}$ is used in (3.93). If we are to accept $\Delta n = \frac{3}{2\ln 2}$, see paragraph 3.1, we shall find $t = \frac{2\ln 2n}{3}\sqrt{\frac{s}{2}} \approx 0.462n\sqrt{\frac{s}{2}}$.

In view of the new independent variable the equation (3.93) will be transformed to the kind:

$$\frac{d^2 q_f}{dt^2} + \varepsilon\frac{dq_f}{dt} = -\left(q_f^2 - q_f^3\right), \tag{3.94}$$

where it is designated $\varepsilon = \dfrac{3}{\sqrt{2s}}$. At variation of a selection standard parameter $0 \le s \le 1$ the size $\infty \ge \varepsilon \ge \dfrac{3}{\sqrt{2}} \approx 2.12$.

The nonlinear differential equation (3.94) can be solved only numerically.

Initial conditions, first of all, are necessary for the solution of the differential equation (3.94): initial frequency of the recessive allele *a* q_{f0} and initial speed of falling of this frequency $\left(\dfrac{dq_f}{dt}\right)_0$ or $\left(\dfrac{dq_f}{dn}\right)_0$.

The initial frequency q_{f0} at $t = 0$ or $n = 0$ as well as in paragraph 3.4 (or Figure 3.4) was accepted $q_{f0} = 0.605$, see also (2.8). The initial speed of falling of the allele frequency we shall accept such that the solution of the equation (9.34) approximately corresponded to a curve 2 on Figure 3.4. In this case $\left(\dfrac{dq_f}{dt}\right)_0 = -0.6$ and $\left(\dfrac{dq_f}{dn}\right)_0 = -0.165$ for the characteristic of hemophilia, the parameter of selection $s = 0.71$.

The step of calculation on a variable *t* equal $\Delta t = 0.001$ and numbers of steps 20000 was used.

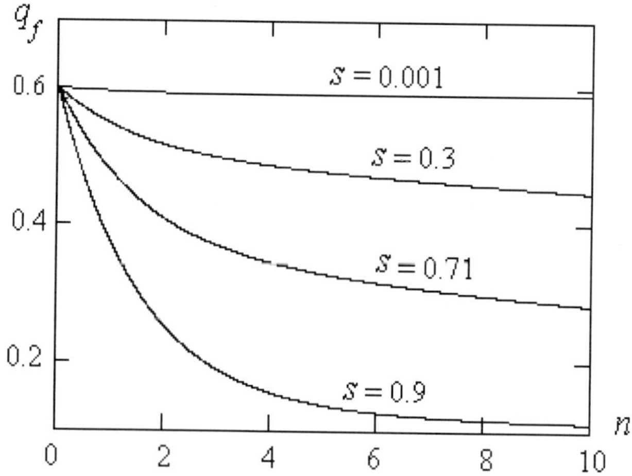

Figure 3.11.

In Figure 3.11 the graphs of dependence of intact recessive allele frequency q_f in *X*-chromosomes in women in a population on dimensionless time *n* are shown at various parameters of selection *s*.

Let us consider the compatible actions of natural selection and inbreeding. Having united the equations (3.79) and (3.93), we shall receive:

$$\frac{d^2 q_f}{dn^2} + \frac{3}{2\Delta n}\frac{dq_f}{dn} = \frac{1}{\Delta n^2}\left(\frac{F}{1-F} - \frac{s}{2}\right) q_f^2 + \frac{s}{2\Delta n^2} q_f^3 . \qquad (3.95)$$

Usually $\frac{F}{1-F} << \frac{s}{2}$, i.e., selection easily compensates for the increase of mutagen action at inbreeding. However, as an example for calculation we shall examine a modeling situation when $\frac{F}{1-F} = \frac{s}{2}$, i.e., weak selection takes place against the recessive homozygotes or power inbreeding in a population. In this case the differential equation (3.95) becomes considerably simpler:

$$\frac{d^2 q_f}{dn^2} + \frac{3}{2\Delta n}\frac{dq_f}{dn} = \frac{s}{2\Delta n^2} q_f^3 . \qquad (3.96)$$

Using the independent variable $t = \frac{n}{\Delta n}\sqrt{\frac{s}{2}}$ applied in the solution of the equation (3.93), we shall find:

$$\frac{d^2 q_f}{dt^2} + \varepsilon \frac{dq_f}{dt} = q_f^3 , \qquad (3.97)$$

as well as $\varepsilon = \frac{3}{\sqrt{2s}}$.

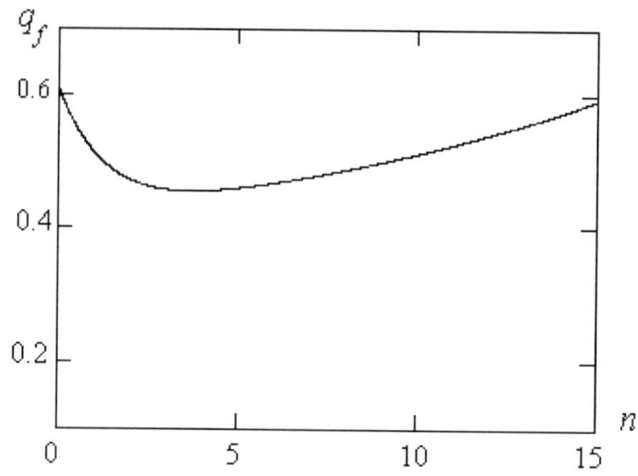

Figure 3.12.

In Figure 3.12 the dependence of the intact recessive allele *a* frequency q_f in *X*-chromosomes in the women in a population on dimensionless time n is shown at parameter of selection $s = 0.71$ (similar for hemophilia) against the recessive homozygotes.

The initial frequency q_{f0} at $n = 0$ was accepted as $q_{f0} = 0.605$. The initial speed of the reduction of the healthy allele *a* frequency was accepted as $\left(\frac{dq_f}{dn}\right)_0 = -0.138$.

From the figure it can be seen that natural selection originally results in a reduction in the frequency of the healthy allele. However, owing to inbreeding its growth then begins at the association of alleles in recessive homozygotes *aa* on account of the very large inbreeding factor $F = \frac{s}{2+s} = 0.262$. Thus, in frequency dependence the minimum of the healthy allele frequency approximately in the fourth generation is observed. More precisely this occurs at the time of a population life which corresponds to an average of four times one generation's alternation, i.e., approximately 100 years.

Let us note that the heir of Tsar Alexei of Russia (who was killed in 1918; Figure 2.8, paragraph 2.3) was ailing from a severe form of hemophilia, having been born 4 generations after Queen Victoria (if we are to assume her generation as being the first) who was born in 1819 and at which point the given mutation *de novo* was observed. Thus, it was the most dangerous to a generation of the family tree of the European royal houses from the point of view of hemophilia, since in the given family tree rather considerable inbreeding also took place.

In conclusion of the given paragraph we shall consider the some general biological questions connected to the functioning of natural selection.

There is a very high degree of nonlinearity - the third - of the differential equation (3.94) describing natural selection in populations. It has great general and biological importance. The degree of nonlinearity of natural selection defines its opportunities, i.e., *power of selection*. We did not succeed in finding out the mutational processes with a higher degree of nonlinearity. The inbred mutational processes have the second degree of nonlinearity.

Other processes resulting, for example, in hemophilia, incomplete inheritance or induced mutagenesis under action of radiation have the first degree of nonlinearity; or, in other words, these processes are linear. This means that any mutational processes in the populations of the Earth can be levelled by natural selection. From this point of view of H. Muller's fear about the danger of the future biological degeneration of mankind owing to an accumulation of a genetic load on account of inbreeding (see paragraph 3.6.1) is likely to be groundless. The damage done to a genome by mutational processes in earthly populations can cause the destruction of a considerable part of individuals, including people. It is extremely unpleasant and it is necessary to resist it in various ways: through the creation of new medical products, protection from ionizing radiations, etc. But as a whole these negative processes do not practically threaten populations, since natural selection gradually restores the genetic norm in a population.

But it is possible to assume a fantastic situation: that some technologically highly-civilized individuals arrive from another planet to the Earth, and are able to operate genic mechanisms. These individuals can start a mutational process in a human population of a fourth (or higher) order of nonlinearity. Now it is difficult to assume a concrete character of such a process. However, it is completely clear that the natural selection generated in earthly conditions could not reset such a mutational influence. Earthly populations would start to die out completely with the faster degree of nonlinearity of the given hypothetical mutational process. And a population with lower technological opportunities of manipulation of genofund would not be capable of interfering with this process.

3.7. Influence of Crossing-Over on a Population

In paragraph 2.1.5 the problem of crossing-over in a family tree was examined. Similarly, the biological essence of the crossing-over, its purposes and task for evolution, have been considered. We shall therefore not dwell on these questions. We shall research the influence of the crossing-over on a population. It has been earlier shown that, for example, a recessive allele frequency in *X*-chromosomes in the following generation of women according to (2.21) equals:

$$q_{f(n+1)} = \frac{1}{2}[(1-\beta)q_{fn} + (1+\alpha)q_{mn}] = \\ = \frac{1}{2}[q_{fn} + q_{mn}] - \frac{1}{2}[\beta q_{fn} - \alpha q_{mn}], \qquad (3.98)$$

where α and β are randomly time-dependent sizes determined by the crossing-over.

Let us transit in (3.98) from a family tree to a population by a standard technique for which we shall transform (3.98), taking into account that the allele frequency in the *X*-chromosome of the woman is equal to the same frequency of alleles in the man of the following generation $q_{mn} = q_{f(n-1)}$. As a result, we find:

$$q_{f(n+1)} = \frac{1}{2}[q_{fn} + q_{f(n-1)}] - \frac{1}{2}[\beta q_{fn} - \alpha q_{f(n-1)}]. \qquad (3.99)$$

Let us transform (3.99), multiplying by -2 to the kind:

$$(1-\beta)q_{fn} - 2q_{f(n+1)} + (1+\alpha)q_{f(n-1)} = 0. \qquad (3.100)$$

Let us consider the differential equation:

$$\frac{d^2 q_{fn}}{dn^2} + \eta \frac{dq_{fn}}{dn} = \mu q_{fn}, \qquad (3.101)$$

where η and μ are random variables.

Let us pass in (3.101) to the finite-differential form:

$$\frac{q_{f(n+1)}-2q_{fn}+q_{f(n-1)}}{\Delta n^2}+\eta\frac{q_{fn}-q_{f(n-1)}}{\Delta n}=\mu q_{fn}. \qquad (3.102)$$

Let us combine in (3.102) similar members and we shall increase the result on -2. As a result we shall find:

$$-2q_{f(n+1)}+2\left(2-\eta\Delta n+\mu\Delta n^2\right)q_{fn}-2\left(1-\eta\Delta n\right)q_{f(n-1)}=0. \qquad (3.103)$$

Let us identify formulas (3.103) and (3.100). As a result we have the system of two algebraic equations:

$$2\left(2-\eta\Delta n+\mu\Delta n^2\right)=1-\beta, \qquad (3.104)$$

$$-2\left(1-\eta\Delta n\right)=1+\alpha. \qquad (3.105)$$

The solution of this system appears thus:

$$\eta=\frac{3+\alpha}{2\Delta n}. \qquad (3.106)$$

$$\mu=\frac{\alpha-\beta}{2\Delta n^2}. \qquad (3.107)$$

Hence, the differential equation (3.101) can be presented in the following way:

$$\frac{d^2q_f}{dn^2}+\left(\frac{3+\alpha}{2\Delta n}\right)\frac{dq_f}{dn}+\frac{\alpha-\beta}{2\Delta n^2}q_f=0. \qquad (3.108)$$

The number of the generation for a population does not play a role – it is a continuous time scale; therefore, an index n at allele frequency is usually omitted.

Taking into account the similarity of the equations (3.108) and (3.62), we conclude that crossing-over is some analogue of incomplete inheritance. The main distinction is that the incomplete inheritance process is determined and crossing-over is stochastic, since α и β in the formula (3.98) random variables.

The equation (3.108) is the linear differential equation. We shall present the allele frequency as $q_f = \overline{q}_f + \Delta q_f$ where $\overline{q}_f$ is some average value of frequency, Δq_f - random deviation of frequency for the account of crossing-over. Substituting the given sum in (3.108) and believing that the average value of frequency $\overline{q}_f$ also satisfies the equation (3.108), we shall find:

$$\frac{d^2\Delta q_f}{dn^2} + \left(\frac{3+\alpha}{2\Delta n}\right)\frac{d\Delta q_f}{dn} + \frac{\alpha-\beta}{2\Delta n^2}\Delta q_f = 0. \quad (3.109)$$

Let us multiply the equation (3.109) on size Δq_f:

$$\Delta q_f \frac{d^2\Delta q_f}{dn^2} + \left(\frac{3+\alpha}{2\Delta n}\right)\Delta q_f \frac{d\Delta q_f}{dn} + \frac{\alpha-\beta}{2\Delta n^2}\Delta q_f^2 = 0. \quad (3.110)$$

Using the identity $\Delta q_f \frac{d^2\Delta q_f}{dn^2} = \frac{d^2}{dn^2}\left(\frac{\Delta q_f^2}{2}\right) - \left(\frac{d\Delta q_f}{dn}\right)^2$ we shall find:

$$\frac{d^2}{dn^2}\left(\frac{\Delta q_f^2}{2}\right) - \left(\frac{d\Delta q_f}{dn}\right)^2 + \left(\frac{3+\alpha}{2\Delta n}\right)\frac{d}{dn}\left(\frac{\Delta q_f^2}{2}\right) + \frac{\alpha-\beta}{2\Delta n^2}\Delta q_f^2 = 0$$

(3.111)

Let us carry out averaging the received equation on dimensionless time n. As a result, we shall receive:

$$\frac{d^2}{dn^2}\left\langle \Delta q_f^2 \right\rangle - 2\left\langle \left(\frac{d\Delta q_f}{dn} \right)^2 \right\rangle + \left\langle \left(\frac{3+\alpha}{2\Delta n} \right) \frac{d\Delta q_f^2}{dn} \right\rangle + \left\langle \frac{\alpha-\beta}{\Delta n^2} \Delta q_f^2 \right\rangle = 0 \tag{3.112}$$

where angular brackets mean the size average on time.

As the size of the allele frequency is limited to a condition $q_f \leq 1$ and the crossing-over process in a population as a whole is stationary, it is possible to assume $\left\langle \Delta q_f^2 \right\rangle = \delta$, where δ is constant. In addition, we shall assume $\alpha << 3$. As a result, we shall find:

$$\left\langle \left(\frac{d\Delta q_f}{dn} \right)^2 \right\rangle = \left\langle \frac{\alpha-\beta}{2\Delta n^2} \Delta q_f^2 \right\rangle. \tag{3.113}$$

Let us introduce the function of a distribution probability density of size Δq_f on dimensionless time n according to determination $f(n) = \frac{1}{\Delta q_f}\frac{d\Delta q_f}{dn}$.

As a result of crossing-over, let the general variation of an allele's number on the average on one X-chromosome during time n of supervision be equal to $N \sim \Delta q_f$. Over time the number of these alleles in the X-chromosome in generations of women on account of crossing-over constantly varies (Figure 2.5). In the time interval the change in the number of alleles is proportional $dN \sim d\Delta q_f$. Hence, the probability $dP = \frac{dN}{N} = \frac{d\Delta q_f}{\Delta q_f}$ for the random variable Δq_f to get in a unit interval of dimensionless time n equals $f(n) = \frac{dP}{dn}$.

Using the function of probability density, it is possible to copy the formula (3.113) as:

$$\left\langle \left(\Delta q_f f(n)\right)^2 \right\rangle = \left\langle \Delta q_f^2 f^2(n) \right\rangle = \left\langle \frac{\alpha-\beta}{2\Delta n^2} \Delta q_f^2 \right\rangle . \tag{3.114}$$

Hence:

$$f(n) = \frac{\sqrt{\langle \alpha-\beta \rangle}}{\Delta n \sqrt{2}} . \tag{3.115}$$

The concrete kind of probability density $f(n)$ of function Δq_f distribution can be defined only from experiment.

Let us initially assume that crossing-over is a passive stochastic process. In this case it is logical to assume that distribution $f(n)$ submits to the normal law [14] as:

$$f(n) = \frac{1}{\sigma\sqrt{2\pi}} e^{-\frac{n^2}{2\sigma^2}} , \tag{3.116}$$

where σ is the mean square deviation of the random variable Δq_f. An obvious limitation $n \geq 0$ is imposed on the argument of function $f(n)$.

Comparing (3.115) and (3.116), we find $\Delta n = \sigma\sqrt{\pi}$ and:

$$\sqrt{\langle \alpha-\beta \rangle} = e^{-\frac{\pi}{2}\left(\frac{n}{\Delta n}\right)^2} . \tag{3.117}$$

Hence, the function of distribution (3.115) appears thus:

$$f(n) = \frac{1}{\Delta n \sqrt{2}} e^{-\frac{\pi}{2}\left(\frac{n}{\Delta n}\right)^2} . \tag{3.118}$$

In Figure 3.11 the dependence $f(n)$, plotted under the formula (3.118) for $\Delta n = \frac{3}{2\ln 2}$ (see paragraph 3.1), is shown.

Apparently from the graph the probability of change owing to the crossing-over of the allele frequency disappears approximately to $n \approx 4$. The received result does not correspond to reality at all. Crossing-over in a population exists constantly and apparently it is genetically caused. The given contradiction specifies that the normal law of distribution (3.116) which is characteristic of passive stochastic systems for the crossing-over is not applicable.

The law of uniform density of probability is more adequate for crossing-over as a genetically supported process; i.e., it is necessary to assume $\sqrt{\langle \alpha - \beta \rangle} = const = \varsigma$. In this case the density of probability of the allele frequency change in generations of a population (3.115) looks like:

$$f(n) = \frac{\varsigma}{\Delta n\sqrt{2}}. \tag{3.119}$$

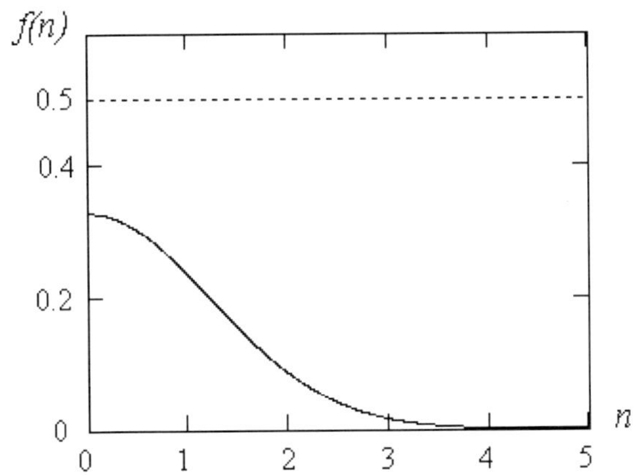

Figure 3.11.

The constant ς can be estimated from the normalizing condition $\int_0^{\infty} f(n)dn = 1$. Believing that in both of the following generations the "intensity" of crossing-over is constant, it is possible to write down a condition normalisation as $\int_0^{2} f(n)dn = 1$. Hence, $\frac{\varsigma}{\Delta n\sqrt{2}} = \frac{1}{2}$ and $\varsigma = \frac{\Delta n}{\sqrt{2}}$, and function $f(n) = \frac{1}{2}$. This level of probability density is also marked on Figure 3.11.

In conclusion, we shall emphasize that the performed analysis shows that crossing-over is not a passive stochastic process. It is a genetically caused, active, biological process. The question of the mechanism of crossing-over genetic regulation remains open.

3.8. Migration of a Population

Populations carry out various motions during vital activity, i.e., migrate in search of the best places to live upon there being an excess of individuals which can exist in the given region; migrate in order to prevent collisions with other populations, in order to gain new territories, etc. The entire history of human population development is connected to the migration of its parts in various regions of the globe.

Migratory processes have laws and genetic consequences. The best known example of the consequences of migratory processes is the change in the human skin pigmentation at the migration of people from Africa where their occurrence is supposedly in higher latitudes which was caused by the adaptation of a population to new conditions of existence. At the same time the process of a population's motion influenced its genome.

Without concerning ourselves with the concrete reasons for migration, we shall analyze genic changes in a migrating population under various conditions.

3.8.1. Migration of Panmictic Populations

Let us begin with the elementary case of the migration of a panmictic population.

The movement of a population of individuals is not equivalent to the movement of genomes. But individuals of the population are carriers of genomes; therefore, the movement of the population of individuals and the movement of genomes are closely connected with each other.

Let us assume that at the migration of the population it retains its panmictic character. Though such an assumption is a significant idealization, for the initial analysis it is nevertheless allowable, especially if the moving part of the population is great enough i.e., contains plenty of family trees.

For the solving of our problem it is necessary to have, first of all, the base equation which describes the genome of the moving population. We shall consider requirements which are necessary to satisfy such an equation.

First of all, this equation should include the Hardy-Weinberg law i.e., to pass the Hardy-Weinberg law in the absence of a population movement.

Second, the equation should reflect the indifferent character of genome balance for moving panmictic populations.

Thirdly, since the movement of a population represents some wave process including elements of individual diffusion in space, it is possible to assume that the equation for moving genomes should have a diffusive-wave character.

For a basis we shall accept the differential equation reflecting the balance of the genome of the motionless panmictic population or the Hardy-Weinberg balance (3.15):

$$\frac{d^2 q_f}{dn^2} + \ln 2 \frac{dq_f}{dn} = 0, \tag{3.120}$$

where q_f is the frequency of some recessive allele of two-allele systems in the *X*-chromosome in the woman, $n = \frac{t}{T}$ - dimensionless time, t - current time, T - the normalizing time equal to the average period of a generation's alternation in the moving population $(T \approx 25\ years)$.

Taking into account the required diffusive-wave character of the equation, we shall write down the base equation for the genome of the moving population as:

$$\frac{d^2 q_f}{dn^2} + \ln 2 \frac{dq_f}{dn} = D \frac{d^2 q_f}{dX^2}, \tag{3.121}$$

where $D = D^* T$ is the normalized factor of diffusion, D^* - the factor of diffusion of the population of individuals, and hence of the genome in space. This factor is multiplied for normalizing time T with the purpose of dimensionless time use in the unit of factor D measurements, X - spatial coordinate of the population moving.

Let us note that the equation (3.121) simultaneously reflects the indifferent character of the genome balance of the moving population since it satisfies the solving $q_f = const$.

The equation (3.121) allows us to analyze the movement of the population and its genome in the certain direction X that most frequently took place in the history of mankind. If the movement of a population occurs simultaneously in different directions it is necessary to use the equation:

$$\frac{d^2 q_f}{dn^2} + \ln 2 \frac{dq_f}{dn} = D \Delta q_f, \tag{3.122}$$

where Δq_f is the Laplacian of the functions q_f.

With the purpose of solving the finding of the equation (3.121) we shall make a replacement of the variable under the formula:

$$q_f(n, X) = e^{-\left(\frac{1}{2}\ln 2\right)n} u(n, X), \tag{3.123}$$

where $u(n, X)$ is the new variable dependent on dimensionless time and coordinate of movement.

Substituting (3.123) in (3.121), we shall receive the equation:

$$\frac{\partial^2 u}{\partial n^2} = D\left(\frac{\partial^2 u}{\partial X^2}\right) + \left(\frac{1}{2}\ln 2\right)^2 u \cdot \tag{3.124}$$

The result of the solution of the Cauchy problem for the given equation has a rather complex character for the analysis. For example, for initial conditions: at $n = 0$ the function $q_f = q_{f0} = u_0 = const$, and $\frac{\partial q_f}{\partial n} = \frac{\partial u_0}{\partial n} = 0$ (initial constant on time value of recessive allele frequency), and also in spatial area $-\infty < X < \infty$ the solution of the equation (3.124) looks like [22]:

$$u = \frac{1}{2}\left[f\left(X + \sqrt{D}n\right) + f\left(X - \sqrt{D}n\right)\right] + \\ + u_0 \frac{n \ln 2}{4\sqrt{D}} \int_{X-\sqrt{D}n}^{X+\sqrt{D}n} \frac{I_1\left(\frac{1}{2}\ln 2\sqrt{n^2 - \frac{(X-\xi)^2}{D}}\right)}{\sqrt{n^2 - \frac{(X-\xi)^2}{D}}} d\xi, \tag{3.125}$$

where f is any function of coordinates $X + \sqrt{D}n$ and $X - \sqrt{D}n$, ξ - variable of integration, $I_1(z)$ - modified Bessel function of the first order.

Function $f\left(X - \sqrt{D}n\right)$ means a wave of the population genome, and hence the wave of the population spread to the right, and $f\left(X + \sqrt{D}n\right)$ - to the left.

Let us consider in more detail the form of a migrating population wave. It is obvious that the approximation of uniform movement in this case is rather rough. The population goes then after a while stops and develops the particular territory. Then, after an exhaustion of resources, the movement renews, etc. In Figure 3.12 the front of a genomic wave of a moving population is shown.

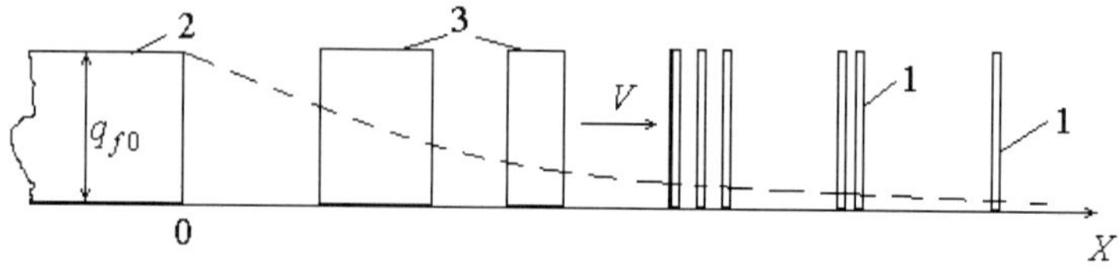

Figure 3.12.

Each man is the carrier of all genomes. Therefore, for the examined allele frequency we shall assume an identical match in each individual of a population $q_f = q_{f0} = const$.

Some individuals, for example, hunters, scouts 1, can leave the basic location of a migrating population 2 on rather big distance. A part of individuals - collectors of a plants fruits 3 leave on smaller distance. In the analysis it is possible to replace such a complex character of the front of some dotted line of a moving population shown in Figure 3.12.

For our purposes it is allowable to use the exponential kind of function *f*. Also we shall examine only a wave of the population genome propagating to the right. In this case the solving of the equation (3.124) can be written down as:

$$u = u_0 e^{-(kX - \omega n)}. \qquad (3.126)$$

In the formula (3.126) we shall be interested the parameters of the wave phase k and ω instead of the wave front which, according to the solution (3.125), could be anyone. Substituting (3.126) in (3.124), we shall find dependence between the parameters of the wave phase and factor of diffusion D:

$$\omega^2 = Dk^2 + \left(\frac{1}{2}\ln 2\right)^2. \qquad (3.127)$$

Substituting (3.126) in (3.123), we shall pass to former function q_f:

$$q_f = q_{f0} e^{-\left(kX - \left(\omega - \frac{1}{2}\ln 2\right)n\right)}. \qquad (3.128)$$

Once again we shall note that no real change of function q_f is present since each individual carries a full genome. Function (3.128) reflects the conditional change shown in Figure 3.12 by the dotted line. In the formula (3.128) $k = \dfrac{2\pi}{\lambda}$ there is the wave number and $\left(\omega - \dfrac{1}{2}\ln 2\right)$ - dimensionless cyclic frequency of the wave population genome, λ - length of the front of the moving population wave. Equality of initial sizes of functions $q_{f0} = u_0 = const$ is also used.

First of all, we shall find the speed of the wave front propagation of the population genome. If the population in the movement keeps its panmictic character the Hardy-Weinberg balance is indifferent and should be observed $q_f = q_{f0} = const$.

Hence, according to (3.128), the phase of the wave is equal:

$$kX - \left(\omega - \frac{1}{2}\ln 2\right)n = 0 \,. \qquad (3.129)$$

The speed of the wave front movement of the population genome, according to (3.129), Figure 3.12, equals:

$$V = \frac{X}{t} = \frac{X}{nT} = \frac{\omega - \frac{1}{2}\ln 2}{kT} \,. \qquad (3.130)$$

Expression (3.130) has a completely general character which has not been connected to the concrete form of a population front wave (3.128).

Substituting in (3.130) parameter ω from (3.127) and taking into account $D = D^{*}T$ we shall find:

$$V = \frac{\sqrt{Dk^2 + \left(\frac{1}{2}\ln 2\right)^2} - \frac{1}{2}\ln 2}{kT} = \sqrt{\frac{D^{*}}{T} + \left(\frac{1}{2kT}\ln 2\right)^2} - \frac{1}{2kT}\ln 2 \qquad (3.131)$$

Let us note that in (3.131) the condition is observed: at $D^* = 0$ the population is motionless $V = 0$.

Using the length of the population wave front under the formula $\lambda = \frac{2\pi}{k}$ we shall find:

$$V = \sqrt{\frac{D^*}{T} + \left(\frac{\lambda}{4\pi T}\ln 2\right)^2} - \frac{\lambda}{4\pi T}\ln 2 \,. \qquad (3.132)$$

It is interesting to analyze on extremum the function of speed $V(T,\lambda)$. Simple transformations of derivatives $\frac{\partial V}{\partial T}$ and $\frac{\partial V}{\partial \lambda}$, and equating of these derivatives with zero, results in a trivial conclusion that the minimal size of speed $V(T,\lambda)=0$ is determined by the condition $D^* = 0$ i.e., a motionless population.

In Figure 3.13 the dependence of the speed of the population genome moving in the average period T of the generation's alternation in the population plotted under formula (3.132) is shown.

For calculation of the length of the population front wave $\lambda = 0.4\, km$ was accepted, the factor of diffusion $D^* = 10\,\frac{km^2}{year}$.

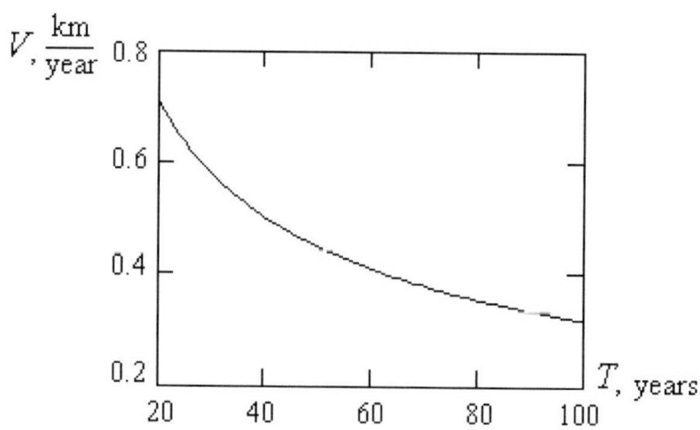

Figure 3.13.

The calculation shows that with the increase in the period of a generation alternation the speed of the population and its genome movement decrease. This is correct since younger individuals are more dynamical i.e., if the generations are are alternated more often, the population move faster. At the period of the generation alternation $T \approx 25\, years$ the calculated speed of genome movement, and hence the population, equals $V \approx 0.63 \dfrac{km}{year}$. It is obvious that this speed is basically caused by a gradual exhaustion of resources on the way of the population movement.

Believing that during migration from Africa primitive people traveled about 5000 *km*, it is possible to find the time of movement ~ 8000 years. For this time 320 generations were alternated.

In Figure 3.14 the dependence of the speed of the population movement on length λ of the wave front is shown at the period of the generation alternation equal $T \approx 25\, years$, plotted under formula (3.132). From the graph it can be seen that with the increase in the length of the wave the speed of the population movement falls. This is correct since more compact populations should move with the greater speed than those which are less compact. But this dependence is not as strong.

At the analysis of the equation (3.132) there is the question of the sense of diffusion factor D^*. This parameter reflects the area developed by the population within a year at each stage of movement. Actually this parameter reflects the resources necessary for vital activity and the movement of the population. We shall notice that the parameter $D = D^*T$ means the area developed by the population during the life of one generation.

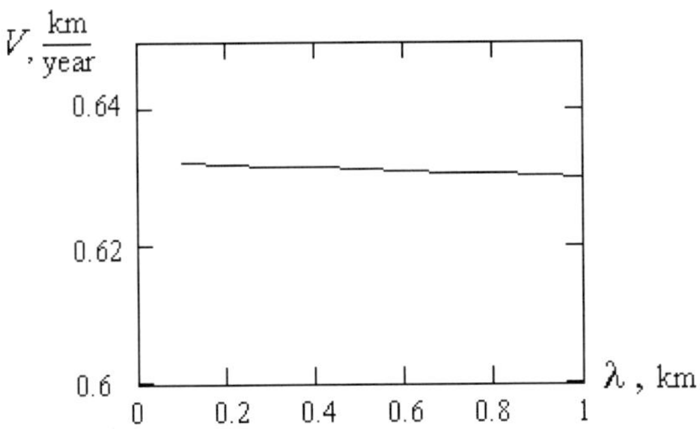

Figure 3.14.

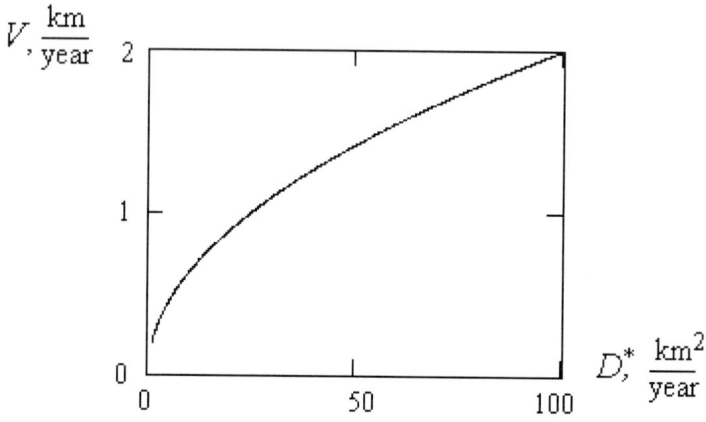

Figure 3.15.

In Figure 3.15 the dependence of the speed of the population movement on diffusion factor D^* is shown.

The found dependence is rather strong. It shows that the bigger the resource area the population can develop for the year of the movement, the faster it can move.

The basic disadvantage of the performed analysis of the moving population is the assumption about its panmictic character. Actually, the migration of the human population part in various directions occurred rather in small groups, tribes or separate communities of tribes. In these tribes there was a relatively small number of family trees and therefore consanguineous mating, i.e., inbreeding was widespread.

3.8.2. Migration of Inbred Populations

As was already specified above in primitive populations, for example, the exodus from Africa and the movement to northern Europe, the influence of inbreeding was important. Therefore, we shall consider the influence of inbreeding on a migrating population.

The way in which the inbreeding accounts for a motionless population has been analyzed in paragraph 3.6. Adding in the equation (3.79) for allele frequency q_f motionless inbred populations the item reflecting the movement of a population (see paragraph 3.8.1) we shall find the general equation for a condition of genome in a moving inbred population:

$$\frac{d^2 q_f}{dn^2} + \frac{3}{2\Delta n}\frac{dq_f}{dn} = D\frac{d^2 q_f}{dX^2} + \frac{F}{\Delta n^2(1-F)}q_f^2, \qquad (3.133)$$

where F is the inbreeding factor.

The found equation is the nonlinear differential equation of the second order with square-law nonlinearity.

We shall search for the solution of the equation (3.133) as a traveling wave $q_f = f(\varsigma)$ where $\varsigma = kX - \omega n$. In this case the equation (3.133) will be transformed to the kind:

$$\frac{d^2 q_f}{d\varsigma^2}\left(\omega^2 - Dk^2\right) - \frac{3}{2\Delta n}\frac{dq_f}{d\varsigma}\omega = \frac{F}{\Delta n^2(1-F)}q_f^2. \qquad (3.134)$$

Let us lower the order of the equation (3.134), having designated $\frac{dq_f}{d\varsigma} = Z\frac{\frac{3}{2\Delta n}\omega}{\omega^2 - Dk^2}$. We shall receive Abel's differential equation:

$$Z\left[\frac{dZ}{dq_f} - 1\right] = \frac{\left(\omega^2 - Dk^2\right)}{\left(\frac{3}{2\Delta n}\right)^2 \omega^2}\frac{F}{\Delta n^2(1-F)}q_f^2 = \frac{4}{9}\left(1 - \frac{D}{V^2}\right)\frac{F}{(1-F)}q_f^2 \qquad (3.135)$$

where $\frac{\omega}{k} = V$ is the speed of the population movement. Let us note that the received differential equation does not depend on size Δn.

The analytical solution of the equation (3.135) does not exist. However, the greatest interest has a dependence of the population movement speed V on inbreeding factor F. This dependence can be found without solving the equation (3.135).

Let us consider the right part of the equation (3.135) which defines the inbreeding in a population. The inbreeding factor F is present both in the numerator, and in the denominator. Factor F serves in the numerator for the establishment of all the existence of the right part in the equation (3.135). At $F=0$ the right part of the equation (3.135) disappears and the population becomes panmictic i.e., in this case $q_f = q_{f0} = const$. If $F \neq 0$ the recessive allele frequency q_f grows (Figure 3.8), i.e., the inbreeding factor in the numerator is responsible for the dynamics of the function q_f growth. Therefore, the inbreeding factor in the numerator of the equation (3.135)'s right part should be kept at any transformations.

Taking into account the given position, we shall transform the right part (3.135) as follows:

$$Z\left[\frac{dZ}{dq_f}-1\right]=\frac{4}{9}\left(\frac{1}{1-F}-\frac{D}{\left(V\sqrt{1-F}\right)^2}\right)Fq_f^2 . \qquad (3.136)$$

All parameters connected to the wave phase are present only at the second addend in brackets of the right part (3.136). Therefore, it is possible to conclude that for the account of the inbreeding factor's influence on the moving population speed it is necessary to multiply the speed of the population wave in $\sqrt{1-F}$. Hence, formulas (3.131) and (3.132) for the genome speed of the moving population and their own population can be transformed thus:

$$V=\left(\frac{\sqrt{Dk^2+\left(\frac{1}{2}\ln 2\right)^2}-\frac{1}{2}\ln 2}{kT}\right)\sqrt{1-F}= . \qquad (3.137)$$

$$=\left(\sqrt{\frac{D^*}{T}+\left(\frac{\lambda}{4\pi T}\ln 2\right)^2}-\frac{\lambda}{4\pi T}\ln 2\right)\sqrt{1-F}$$

In Figure 3.16 the dependence of the moving population's speed on the inbreeding factor is shown at the following parameters: $\lambda = 0.4\,km$, $D^* = 10\,\dfrac{km^2}{year}$, $T \approx 25\ years$.

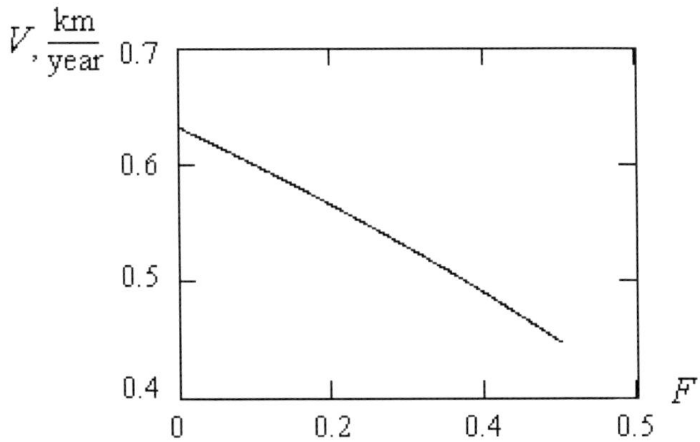

Figure 3.16.

The received result (Figure 3.16) can be interpreted as follows. At an increase of inbreeding in the moving population its speed of movement decreases. Biologically, it can be connected by the fact that in the population with consanguineous mating the individuals become more weakened and less dynamical during moving.

At $F = 1$ according to (3.137) the population stops. The reason for this stop is the fast and full degeneration of the population. If we were to address the equation (3.79) for the motionless population (or (3.133) at $D = 0$) at $F \to 1$ the right part of the equation which determines the inbreeding quickly grows, aspiring to infinity. This increase can be compensated only to an increase in velocity of frequency q_f growth in the left part of the equation (3.79). But this speed is limited by the biologo-reproductive opportunities of the population. Therefore, in a reality achievement $F = 1$ does not occur because the population quickly degenerates.

Thus, the migration of human populations in various regions of the globe is caused by the movement of a population as individuals and therefore the movement of a population genome which is transferred with the individuals in question. The found differential equation for moving the genome of panmictic populations (3.121) allows us to find the dependence of the speed of a moving population on various parameters; in particular, the average time of a generation alternation, the area developed by a population for a certain time, and the length of a wave of a moving population.

During the migration of populations owing to the small number of individuals the role of consanguineous mating was great. The found differential equation (3.133) of the moving inbred populations allows us to analyze the reduction in speed of a population movement depending on the inbreeding factor.

3.8.3. Migration of Inbred Populations in the Presence of Selection

The population is usually in the process of migration for a long enough time. During the migration of a population there is an alternation of generations and natural selection operates.

Using (3.133) and (3.93), it is possible to write down the differential equation migrating inbred populations in which natural selection operates.

$$\frac{d^2 q_f}{dn^2}+\frac{3}{2\Delta n}\frac{dq_f}{dn}=D\frac{d^2 q_f}{dX^2}+\frac{F}{\Delta n^2(1-F)}q_f^2-\frac{s}{2\Delta n^2}\left(q_f^2-q_f^3\right) \quad (3.138)$$

The received nonlinear equation is too difficult for the analysis. However, taking into account the cubic nonlinearity of the differential equation of the second order, it is possible to assume the presence of solutions as solitary population waves [25]. Apparently, the migrating population represents such a solitary wave.

However, the equation (3.138) allows us to receive a very important analytical result. For this purpose wave substitution $\varsigma=\frac{1}{\Delta n^2}(kX-\omega n)$ is used, where k is the wave number, and ω - the dimensionless cyclic frequency of the population waves. As a result, we shall find:

$$\frac{d^2 q_f}{d\varsigma^2}\left(\omega^2-Dk^2\right)-\frac{3}{2}\frac{dq_f}{d\varsigma}\omega=\left(\frac{F}{1-F}-\frac{s}{2}\right)q_f^2+\frac{s}{2}q_f^3. \quad (3.139)$$

Let us find the particular solution of the equation (3.139). We shall equate the right part of this equation to zero $\left(\frac{F}{1-F}-\frac{s}{2}\right)q_f^2+\frac{s}{2}q_f^3=0$. As a result, we have:

$$q_f=\left(1-\frac{2F}{(1-F)s}\right)=const\,. \tag{3.140}$$

Obviously, the value of frequency q_f (or $p_f=1-q_f=\frac{2F}{(1-F)s}$) in the formula (3.140) satisfies the equations (3.138) and (3.139).

The received expression (3.140) can be interpreted as follows. If in inbred populations which began to migrate there is some frequency of a recessive allele in *X*-chromosomes in women for the preservation of this frequency there should be an interrelationship (3.140) between inbreeding factor *F*, parameter of selection *s* and this frequency q_f. Formula (3.140) is correct for motionless populations.

We assume the frequency of the recessive allele of blood O equals $q_f=0.605$; see formula (2.8). If we were to use inbreeding factor $F=19\cdot10^{-5}$, for example, for Germany [3], for which migrating population this frequency did not vary, the parameter of selection must be $s=9.6\cdot10^{-4}$.

The problem of interaction of populations is very important. Obviously, it is connected to the finding of the multiwave solutions of the equation (3.138) which is an extremely complex mathematical problem. Population waves can cooperate with each other in various ways. For example, they can destroy each other as interacting nervous impulses [26]. Obviously, conflicted interactions of migrating populations correspond to this.

Population waves can have the character of solitons [25].

Solitons are stable nonlinear solitary waves which show particle-like properties at interaction with other local perturbations or with each other. For example, they restore their initial form, i.e., interactions similar to absolutely elastic particles, for example, to absolutely elastic spheres. As well as spheres, the solitons, for example, can be reflected from a barrier. The marked properties of solitons are caused by the strict balance of the nonlinear processes resulting in an increase of a steepness of a wave forward front and dispersive processes that lead to the destruction of a solitary wave, i.e., to the reduction of a front steepness. The balance of these processes leads to stability of the soliton, i.e., invariability of its form.

We shall note that the third degree of nonlinearity of the differential equation (3.138) is sufficient for the occurrence of a soliton [25]. At interaction of the solitons the population waves can change directions on opposite by analogy impacting spheres; i.e., populations try to avoid the destruction. If populations belong to poorly cooperating individuals (elephants and birds), waves of populations at interaction pass through each other practically not influencing one another.

As a whole, the further analysis of interaction of populations has no genetic-mathematical character and passes to the area of social phenomena.

Chapter 4

CONCLUSION

This book can be attributed to the area of population genetics. The book is practically divided into two parts: mathematical genetics of a family tree and mathematical genetics of populations. These two parts are interconnected and overlap one another in certain methods.

After reading the given book the base of the order statements of the book material is clear to the reader. First, the terminology of the analyzed processes is determined. Then some basic phenomena whose biological characters are included in the consideration: the process of duplication, mating, inheritance, etc. But in the center of the examined problems undoubtedly there is the Hardy-Weinberg law.

Found simultaneously by the English mathematician Hardy and the German physician Weinberg in 1908 on the basis of earlier experiments (1865) which were carried out by the Austrian (Czech) scientist Gregor Mendel, this law is the basic tool of the analysis.

The wonderful feature of this analysis is the reception of rather numerous and important new results without manipulation of the molecular bases of genetics. These results, first of all, are connected to a new method of the separate approach to a family tree and a population which was developed by the author. The rather fruitful method shown has allowed us to receive quantitative results in many problems in which the population genetics were applied earlier to a limited extent; for example, in the radiating influence on a population, migration of populations, problems of oncology, etc.

From all the material of the book it is clear that population genetics are far from being a completed part of general genetics. In particular, the problem of the interaction of populations has not been mentioned for analysis until now.

Therefore, there are still more many interesting achievements and openings ahead.

REFERENCES

[1] Libbert E. "Kompendium der allgemeinen biologie." VEB Gustav Fisher Verlag, Jena. 1982. 440 p.

[2] Bernstein F. "Zusammenfassende Betrachtungen uber die erblichen Blutstrukturen des Menschen." (1925), Z. Induktive Abstammungs-Vererbungslehre, 37, p. 237.

[3] Vogel F., Motulsky A. "Human genetics." Volumes 1, 2, 3, Springer-Verlag, Berlin, 1990. 1068 p.

[4] Holdane J.B.S. "The rate of spontaneous mutation of a human gene." (1935), *J. Genet.*, 31, pp. 317-326.

[5] Danforth G.H. "The frequency of mutation and the incidence of hereditary traits in man. Eugenics genetics and the family." (1921), *Scientific Paper of the 2nd International Congress of Eugenics*, Vol. I, New York, pp. 120-128.

[6] Li, C.C. "*First Course in Population Genetics.*" The Boxwood Press Pacific Grove, California, 1976. 556 p.

[7] Ayala F., Kiger, J. Jr. "*Modern genetics.*" Volume 3, The Benjamin/Cummings Publishing Company, Inc. California, 1984. 335 p.

[8] Yaworsky B.M., Detlaf A.A. "*Handbook on Physics.*" Moscow, Science, 1990. 624 p.

[9] Kostuk P.G., Grodzinsky D.M., et al. "*Biophysics.*" Kiev, Higher School, 1988. 504 p.

[10] Godunov S.K., Rjabenky V.S. "*Differences scheme.*" Moscow, Science, 1977. 440 p.

[11] Wright S. "Evolution in Mendelian population." (1931), *Genetics.* 19. pp. 97-159.

[12] Muller H.J. "Our load of mutation." (1950), *Am. J. Hum. Genet.* 2, pp. 111-176.

[13] Jimulev I.F. "*General and molecular genetics.*" Novosibirsk, Siberian University Publishing House, 2007. 479 p.

[14] Ventcel E.S. "Probability theory." Moscow, *Science,* 1969. 576 p.

[15] Trapeznikov N.N., Shain A.A. *"Oncology."* Moscow, Medicine, 1992. 400 p.

[16] Matveev A.N. "*Molecular physics.*" Moscow, Higher School, 1981. 400 p.

[17] Kalabekov I.G. "*Russian Reforms in Digits and Facts.*" Moscow, RUSAKI, 2010. 498 p.

[18] Kudryashov J.B. "Radiating biophysics." Moscow, *Fizmatlit,* 2004. 448 p.

[19] Volobuev A.N. "*Bases of Medical and Biological Physics.*" Moscow, Samara Printing House, 2011. 672 p.

[20] Volobuev, A.N., Romanchuk P.I., & Malishev V.K. "Nonlinear Genetics. Inbreeding and Genetic Load." (2014), Canada, Advances in *Natural Science*, 7(1), pp. 1-5.

[21] Morton N.E., Crow J.F., Muller H.J. "An estimate of the mutational damage in man from data on consanguineous marriages." (1956), Proc. *Natl. Acad. Sci. USA,* 42, pp. 855-863.

[22] Zajtsev V.F., Polanin F.D. "Handbook on the differential equations with particular derivatives: accuracy solutions." Moscow, *International Program of Education,* 1996. 496 p.

[23] Brown D., Rothery P. "*Models in Biology: Mathematics, Statistics, and Computing.*" Chichester, NY, Jon Wiley & Sons Ltd., 1994. 688 p.

[24] Carola R., Harley J.P., Novack C.R. "*Human Anatomy and Physiology.*" McGraw-Hill Publishing Company, New York, San Francisco, London, 1990. 926 p.

[25] Dodd R., Eilbek G., Ghibbon G., Morris G. "*Solitons and Nonlinear Wave Equations.*" Moscow, World, 1988. 696 p.

[26] Volobuev A.N., Neganov V.A., Zajtsev V.V. "Physical and mathematical nature of the nervous pulse." (1998) *Physics of wave processes and radio engineering systems,* 1, 2-3, 108-110.

INDEX

E

F

G

H

I

J

K

L

M

N

O

P

R

S